HANDBOOK OF ANTIOXIDANTS

Bond Dissociation Energies, Rate Constants, Activation Energies and Enthalpies of Reactions

Evguenii Denisov

CRC Press is an imprint of the
Taylor & Francis Group, an informa business

First published 1995 by CRC Press
Taylor & Francis Group
6000 Broken Sound Parkway NW, Suite 300
Boca Raton, FL 33487-2742

Reissued 2018 by CRC Press

Library of Congress Cataloging-in-Publication Data

Denisov, E.T. (Evguenii Timofeevich)
Handbook of antioxidants : bond dissociation energies, rate constants, activation energies, and enthalpies of reactions / Evguenii T. Denisov.
p. cm.
Includes bibliographical references and index.
ISBN 0-8493-9426-0 (alk. paper)
1. Antioxidants—Handbooks, manuals, etc. I. Title.
QD281.O9D45 1995
547'.23—dc20 95-21651

A Library of Congress record exists under LC control number: 95021651

ISBN 13: 978-1-315-89327-3 (hbk)
ISBN 13: 978-1-351-07237-3 (ebk)

Visit the Taylor & Francis Web site at http://www.taylorandfrancis.com and the
CRC Press Web site at http://www.crcpress.com

I dedicate this handbook to the memory
of Viktor Kondratev, who inspired me
to work in the field of
quantitative kinetic information.

PREFACE

The objective of this Handbook is to provide scientific workers and engineers working in the field of physical chemistry of antioxidants with the comprehensive data on the bond dissociation energies of reactions flowing in oxidizing substances in the presence of inhibitors.

Autoxidation of hydrocarbons and other organic compounds is a reaction of unique importance for mankind and all living organisms on the earth. Oxidation of organic compounds is one of the important routes of organic synthesis in modern chemical industry. At the same time autoxidation is detrimental in some cases. Keeping and using various chemical products in air often results in their rapid deterioration. There are many such products, including fuels, lubricant oils, rubber, polymers, chemicals, solvents, food stuffs, etc. For this reason, a very important branch of applied science is the chemistry of antioxidants.

During the last 30 years free radical reactions of antioxidants were intensively studied by different kinetics methods, and rate constants of hundreds of reactions of peroxyl radicals with phenols, amines etc. were measured. These data constitute the basic ground for kinetic analysis of systems of the type RH + O_2 + antioxidants. However many important reactions of antioxidants and their intermediates have not been characterized with kinetic parameters. This problem may be solved by using semiempirical methods of rate constants evaluation. The parabolic model of transition state of free radical reaction was chosen in this Handbook to solve this problem. So the Handbook contains tables with experimentally measured rate constants as well as with those calculated by formulas of parabolic model.

The following data are collected in this Handbook: Bond dissociation energies of antioxidants such as phenols (O—H-bonds), aromatic amines (N—H-bonds), hydroxyl amines (O—H-bonds), thiophenols (S—H-bonds); activation energies and rate constants of reactions of peroxyl radicals with antioxidants; rate constants of reaction of phenoxyl, aminyl and nitroxyl radicals with RH, ROOH, phenols, thiophenols, amines and hydroxyl amines; rate constants of reactions of antioxidants with hydroperoxides and oxygen. All information on RH oxidation and antioxidants is divided in to 7 chapters. The first is devoted to short description of mechanism and kinetics of autoxidation of hydrocarbons in the presence of antioxidants , including mechanisms of cyclic chain termination by some inhibitors. It contains the description of the parabolic model of transition state and main formulas for rate constants calculation. The values of rate constants of elementary steps of hydrocarbon autoxidation as well as that of initiators decay. are given in the second chapter. Chapters 3–7 include the thermodynamic and kinetic parameters of reactions of phenols, aromatic amines, hydroxylamines, thiophenols, thiocarbamates and thiophosphates, that are involved in oxidation of hydrocarbons with these compounds.

Symbols and units used in Handbook are in accordance with IUPAC recommendation written in the manual, Quantities Units and Symbols in Physical Chemistry, Blackwell Scientific Publications, London, 1988.

All comments, critical notes, and suggestions will be welcomed by the author. Address to send comments to author is the following: Institute of Chemical Physics, Chernogolovka, Moscow Region, 142432, Russia and E-mail: DENISOV@ICP.AC.RU

I especially thank Taissa G. Denisova for her very valuable help in preparing this manuscript.

I am indebted to Vladimir E. Denisov and Sergey V. Foraponov for their help and advice on Microsoft Word Windows.

Finally I am grateful to Lyudmila N. Pilipetskaya for her rapid and accurate typing.

Chernogolovka, Moscow Region **Evguenii T. Denisov**
June 19, 1995

CONTENTS

LIST OF CHEMICAL SYMBOLS

AmH	aromatic amine
$Am^{\bullet}$	aminyl radical
AmOH	organic hydroxyl amine
$AmO^{\bullet}$	nitroxyl radical
Ar_1OH	phenol
$Ar_1O^{\bullet}$	phenoxyl radical
Ar_2OH	sterically hindered phenol
$Ar_2O^{\bullet}$	sterically hindered phenoxyl radical
ArSH	thiophenol
$ArS^{\bullet}$	thiophenoxyl radical
InH	inhibitor
RH	oxidizing substance
R_1H	aliphatic or alicyclic hydrocarbon
R_2H	olefin hydrocarbon
R_3H	alkylaromatic hydrocarbon
ROOH	hydroperoxide
$RO_2^{\bullet}$	peroxyl radical
RSH	mercaptane

LIST OF PHYSICO-CHEMICAL SYMBOLS

A	preexponential factor of rate constant of reaction in its Arrhenius form, expressed in s^{-1} for unimolecular, in $l\ mol^{-1}\ s^{-1}$ for bimolecular and in $l^2\ mol^{-2}\ s^{-1}$ for trimolecular reaction
b_i	coefficient proportionality between potential energy and amplitude of vibration of i-th chemical bond, $b_i = \pi \nu_i (2\ \mu_i)^{1/2}$
D	bond dissociation energy, expressed in $kJ\ mol^{-1}$
E	activation energy of reaction, expressed in $kJ\ mol^{-1}$
e	the probability of radical pair to go out of cage in liquid phase
f	the stoichiometric coefficient of chain termination by inhibitor
ΔH	enthalpy of chemical reaction, expressed in $kJ\ mol^{-1}$
h	Planck's constant, $h = 6.626 \times 10^{-34}$ J s
K	equilibrium constant
k	rate constant of reaction, expressed in s^{-1} for unimolecular, in $l\ mol^{-1}\ s^{-1}$ for bimolecular and in $l^2\ mol^{-2}\ s^{-1}$ for trimolecular reaction
L	Avogadro's constant, $L = 6.022 \times 10^{23}\ mol^{-1}$
R	gas constant, $R = 8.314\ J\ K^{-1}\ mol^{-1}$
T	absolute temperature, expressed in K
v	rate of chemical reaction, expressed in $mol\ l^{-1}\ s^{-1}$
α	ratio of b coefficients of dissociating and forming bonds in free radical reaction of H-atom abstraction

List of Physico-chemical Symbols

β	ratio of rate constants of hydroperoxide decomposition in to free radicals (k_3) and any products (k_d), $\beta = k_3 / k_d$
θ	$\theta = 2.3RT$ / kJ mol^{-1}
μ	reduced mass of two atoms forming the bond
ν	length of the chain in chain reaction
ν_i	frequency of vibration of i-th bond in molecule
τ	induction period of inhibited oxidation

Chapter 1

KINETICS AND MECHANISM OF INHIBITED OXIDATION OF HYDROCARBONS

1.1 Mechanism of autoxidation of hydrocarbons

The oxidation kinetics and mechanisms of the simplest, though extremely important compounds, hydrocarbons (RH), have been studied in much detail: see, e.g.[1-6]

The early stage of oxidation of an organic compound at the C—H bond resulting in formation of the hydroperoxide consists of the following elementary steps.

$$\begin{array}{rcll}
\text{I (initiator)} & \rightarrow & 2r^{\bullet} & \text{(i)} \\
r^{\bullet} + RH & \rightarrow & rH + R^{\bullet} & \text{(i')} \\
R^{\bullet} + O_2 & \rightarrow & RO_2^{\bullet} & (1) \\
RO_2^{\bullet} + RH & \rightarrow & ROOH + R^{\bullet} & (2) \\
ROOH & \rightarrow & RO^{\bullet} + HO^{\bullet} & (3) \\
ROOH + RH & \rightarrow & RO^{\bullet} + H_2O + R^{\bullet} & (3') \\
ROOH + CH_2{=}CHX & \rightarrow & RO^{\bullet} + HOCH_2C^{\bullet}HX & (3'') \\
2\,ROOH & \rightarrow & RO_2^{\bullet} + H_2O + RO^{\bullet} & (3''') \\
R^{\bullet} + R^{\bullet} & \rightarrow & RR \text{ (or } RH + \text{olefin)} & (4) \\
R^{\bullet} + RO_2^{\bullet} & \rightarrow & ROOR & (5) \\
RO_2^{\bullet} + RO_2^{\bullet} & \rightarrow & ROH + O_2 + R'{=}O \text{ or } ROOR + O_2 & (6)
\end{array}$$

Oxidation is a chain process, if chain propagation reactions 1 and 2 are faster than chain termination reactions 4–6. Radical $R^{\bullet}$ reacts violently with oxygen, $k_1 = 10^7–10^9$ l mol^{-1} s^{-1}; therefore, for oxygen concentrations above 10^{-4} M, $[RO_2^{\bullet}] >> [R^{\bullet}]$ and the chains are terminated by reaction 6. Under these conditions oxidation rate

$$v = v_i + k_2 (2k_6)^{-1/2} [RH] v_i^{1/2} \quad (1.1)$$

and chain mechanism takes place when initiation rate $v_i < 0.5\, k_2^2\, k_6^{-1}\, [RH]^2$.

Quasistationary concentration of peroxyl radicals $[RO_2^{\bullet}]_s$ in oxidizing substance RH is reached in period of time $\tau = 0.74(2k_6 v_i)^{-1/2}$ and usually varies from 0.1 to 10^2 s. When initiator I is the main generator of chains, the rate of oxidation at long chains proceeds with constant rate, if the decay of initiator is negligible, and oxygen is consumed by oxidizing RH with constant rate

$$\Delta[O_2] = k_2 (2k_6)^{-1/2} [RH] (k_i [I]_0)^{1/2} t \quad (1.2)$$

When oxidation is proceeding during a long period of time so that the initiator concentration is decreasing and the rate of free radicals generation is changing along the experiment, the kinetics of oxygen consumption $\Delta[O_2](t)$ is described by the following formula

$$\Delta[O_2] = a(1 - e^{-0.5kt}) \quad (1.3)$$

where k is the rate constant of initiator decomposition, $a = 2k_2(k_i/k_6)^{1/2}k^{-1}$ [RH] $[I]_0^{1/2}$.

Autoxidation of hydrocarbons without initiator proceeds with acceleration, that is the result of hydroperoxide formation and its decomposition into free radicals. The chain generation in the initial period of autoxidation proceeds via a very slow reaction RH with oxygen (see Chapter 2). Due to hydroperoxide formation the rate of chain generation is increasing with time and subsequently is increasing the rate of oxidation v

$$v = k_2(2k_6)^{-1/2}[RH]\,(v_{i0} + k_3\,[ROOH])^{1/2} \qquad (1.4)$$

where $v_{i\,0}$ is the rate of free radicals formation by reaction RH with O_2. Also hydroperoxide is decomposing slowly at its low concentration, so that it is nearly equal to consumed oxygen during some period of time ($t << k_d^{-1}$, k_d — rate constant of ROOH decay), and at such conditions the kinetics of oxygen consumption is described by the following formula

$$\Delta[O_2] = a\,v_{i0}^{1/2}t + 0.25a^2k_3t^2 \qquad (1.5)$$

where $a = k_2(2k_6)^{-1/2}$ [RH].

At very low values of v_{i0} the kinetics of oxygen consumption has the following simple form:

$$(\Delta[O_2]\,)^{1/2} = 0.5\,ak_3^{1/2}t \qquad (1.6)$$

The rate of autoxidation is growing up to the moment when hydroperoxide concentration become quasistationary. Beginning at this moment its concentration [ROOH] = $[ROOH]_s$ and the rate of oxidation is proportional to square of [RH]

$$v = k_2^{\,2}\,(2k_6\,)^{-1}\,(k_3\,/k_d\,)[RH]^2 \qquad (1.7)$$

and is decreasing in time due to oxidation of RH.

1.2. Mechanism of hydrocarbon oxidation inhibited by acceptors of peroxyl radicals

The compounds that inhibit oxidation of hydrocarbons in the liquid phase may be broken up into four groups as regards the mechanism of such inhibition: (1) inhibitors that terminate chains through reactions with peroxyl radicals, including phenols, aromatic amines, hydroxylamines, thiophenols, and aminophenols; (2) inhibitors that terminate chains through reactions with alkyl radicals, including stable radicals, quinones, quinone imines, methylenequinones, nitrocompounds, and condensed aromatic hydrocarbons (these inhibitors are effective when dissolved oxygen concentration is low); (3) agents that decompose peroxides without generating free radicals, including sulfides, disulfides, phosphites, metal thiophosphates, and carbamates; (4) complexing agents that deactivate heavy metals capable of catalyzing hydroperoxide decomposition into free radicals and thereby promoting oxidation, including diamines, amino acids, hydroxy acids, and other bifunctional compounds.

The following reactions take place in the system upon introduction of InH, which reacts with $RO_2^\bullet$.[2,7–14]

$$RO_2^\bullet + InH \rightarrow ROOH + In^\bullet \quad (7)$$
$$In^\bullet + ROOH \rightarrow InH + RO_2^\bullet \quad (-7)$$
$$RO_2^\bullet + In^\bullet \rightarrow \text{products (InOOR)} \quad (8)$$
$$In^\bullet + In^\bullet \rightarrow \text{products} \quad (9)$$
$$In^\bullet + RH \rightarrow InH + R^\bullet \quad (10)$$
$$InH + ROOH \rightarrow \text{products} \quad (11)$$
$$InH + O_2 \rightarrow In^\bullet + HO_2^\bullet \quad (12)$$
$$InOOR \rightarrow InO^\bullet + RO^\bullet \quad (13)$$
$$In^\bullet \rightarrow Q + r^\bullet \quad (14)$$
$$In^\bullet + O_2 \rightarrow Q + HO_2^\bullet \quad (15)$$

Reactions i, 1–15, comprise the principal kinetic scheme of inhibited hydrocarbon oxidation. In a given system these reactions take place with different intensities, so some are not as significant as others.

If InH is so active and its concentration so high that $RO_2^\bullet$ will react faster with InH than with RH, the oxidation will be a non chain radical reaction with rate

$$v = v_i + k_2 [RH][RO_2^\bullet] \quad (1.8)$$

Often, one meets with a different situation: the oxidation is a chain reaction. If $In^\bullet$ takes no part in chain propagation but only reacts with $RO_2^\bullet$, the oxidation rate is

$$v = v_i + k_2 [RH][RO_2^\bullet] = v_i + v_i k_2 [RH] / fk_7[InH] \quad (1.9)$$

As more InH is consumed, the oxidation rate will increase and oxygen absorption kinetics will be given by (for current time $t < f[InH]_o/v_i$)

$$\Delta[O_2] = -(k_2 [RH] / k_7)\ln(1 - v_i t / f[InH]_o) \quad (1.10)$$

$$\Delta([O_2])^{-1} = (k_2 [RH] v_i / fk_7 [InH]_0)^{-1} \Delta(t^{-1}) \quad (1.11)$$

Since $RO_2^\bullet$ is involved in the oxidation and InH consuming reactions, the amount of absorbed oxygen and the amount of consumed InH in chain oxidation of RH are related by

$$\Delta[O_2] = (k_2 [RH] / k_7) \ln([InH]_o/[InH]) \quad (1.12)$$

The $ArO^\bullet$ formed from phenol (ArOH) or the $Am^\bullet$ formed from amine (AmH) may also participate in chain propagation. As we have already pointed out, this may happen via one of four reactions.

The first route is reaction 10. When its rate exceeds that of $In^\bullet$ decay, the kinetics of inhibited oxidation is changed. At $k_{10} [RH][In^\bullet] = 0.5v_i$, the quasi-steady concentration $[RO_2^\bullet] = (k_{10} [RH] v_i / fk_7 k_8 [InH])^{1/2}$ and the rate of chain oxidation is given in Table 1.1. There is given the formula for kinetics of oxygen absorption during the early stages of oxidation also.

The second route may be reaction of $In^\bullet$ with ROOH, reaction −7. In this reaction $Am^\bullet$ and $ArO^\bullet$, which have no *tert*-alkyl substituents in the ortho position (see earlier), actively participate. If this exchange is fast enough (k_{-7} [ROOH][$In^\bullet$] > v_i), it will affect the inhibited oxidation kinetics. The appropriate formulas for absorption kinetics are given in Table 1.1.

Table 1.1
Formulas for the rate v and kinetics of oxygen consumption $\Delta[O_2]$ (t) of inhibited oxidation of hydrocarbons

Limiting steps	v/ mol l^{-1} s^{-1}	$\Delta[O_2](t)$ / mol l^{-1}
2, 7	$v_i(1 + k_2[RH]/fk_7[InH])$	$v_i t - k_2 k_7^{-1}$ [RH] $\ln(1 - t/\tau)$
2, 7, 8, 10	$2k_2[RH]^{3/2}(k_{10} v_i / fk_7 k_8 [InH])^{1/2}$	$2v_0\tau F(t)$
2, 7, −7, 8	$k_2[RH](k_{-7} v_i [ROOH]/fk_7 k_8[InH])^{1/2}$	$v_0^2 \tau^2 [ROOH]_0^{-1} F(t) + 2v_0 \tau F(t)$
2, 7, 11	$k_2[RH](v_i + 2k_{11}[ROOH][InH])/fk_7[InH]$	$2v_0\tau F(t)$
2, 7, 12	$k_2[RH](v_i + 2k_{12}[O_2][InH])/fk_7[InH]$	$2v_0\tau F(t)$
2, 7, 13	$3k_2 v_i [RH]/fk_7[InH]$	$-k_2 k_7^{-1}$ [RH] $\ln(1 - t/\tau)$
2, 7, 14	$k_2[RH](k_{14} v_i / fk_7 k_8 [InH])^{1/2}$	$2v_0\tau F(t)$
2, 7, 9, 15	$k_{15}(2k_9 v_i)^{-1/2}[O_2](2 + k_2[RH]/k_7[InH])$	$a t - a k_2 k_7^{-1}$ [RH]

$\tau = f[InH]_0/v_i$; $F(t) = 1 - (1 - t/\tau)^{1/2}$; $v_0 = k_2[RH]/fk_7[InH]_0$.

The third route of chain propagation may be the breakdown of the inhibitor radical to a radical capable of participating in chain propagation. For example, the *p*-alkoxyphenoxyl radical decomposes to quinone and alkyl radical (reaction 14). On reacting with oxygen the alkyl radical is converted to $RO_2^\bullet$, which propagates chains. As before, an intensive enough decomposition will give rise to a chain process with chains terminated through reaction 8 of the scheme. As seen from Table 1.1, relationships of the type $v \sim v_i^{1/2}$ and $v \sim [InH]^{-1/2}$ are typical of all inhibited oxidation mechanisms with $In^\bullet$ participating in chain propagation. If $In^\bullet$ does not participate, $v \sim v_i$ and $v \sim [InH]^{-1}$.

And the fourth route of chain propagation is the reaction of semiquinone radical with molecular oxygen (reaction 15). This mechanism takes place when diatomic phenols and amines are used as inhibitors.

When trialkylphenols are used to inhibit RH oxidation via the reaction between $RO_2^\bullet$ and $ArO^\bullet$, *o*-and *p*-quinolide peroxides are formed.[11] The quinolide peroxides decompose unimoleculary in hydrocarbons. *o*-Quinolide peroxides decompose much faster than *para*-counterparts, owing primarily to the lower activation energy for decomposition. Investigation of cumene oxidation with quinolide peroxide as initiator has shown that these peroxides decompose to yield free radicals that initiate oxidation. The initiation rate constant $k_i = 2ek_{13}$, where e is the probability for radicals to escape from a solvent cage; in hydrocarbon solutions $e \approx 1$. The degree of decomposition of the quinolide peroxide to radicals will determine the ArOH consumption rate in RH undergoing oxidation. At a temperature where the peroxide is stable, ArOH is consumed with a constant rate v_i = const, and $[InH] = [InH]_o - v_i f^{-1} t$ at a higher temperature ArOH will be consumed with self-acceleration. If the experimental temperature is high enough so that $k_{13}^{-1} << \tau$, the (RH–InH–initiator–O_2) system will within time $t \sim k_{13}^{-1}$ relax to a quasi-steady regime with respect to the peroxide with the total initiation rate $v_{i\Sigma} = v_i + 2ek_{13}$ [InOOR] $= 2v_i\ (2 - e)^{-1}$ if $f = 2$, as is usual for ArOH. The induction period in such a system will be $\tau = 1/2[\text{InH}]_0 v_i^{-1}(2 - e)$; that is decomposition of quinolide peroxide shortens the induction period of phenol-inhibited oxidation.

In hydrocarbon systems ArOH and AmH react not only with $RO_2^\bullet$ but also with oxidants such as O_2 and ROOH (reaction 12). Reaction of inhibitor with oxygen is endothermic since in most cases bond dissociation energy (D) $D_{\text{In—H}} > D_{\text{H—O}_2\bullet}$ and its activation energy is about equal to its endothermicity. Phenols ArOH are oxidized with ROOH slowly (see Chapter 3). The reaction apparently involves abstraction of H atom from ArOH by ROOH and scission of the O—O-bond. The reaction is preceded by H bonding between ArOH and ROOH. As a result of reaction 11 free radicals ($RO^\bullet$) are formed that initiate oxidation. The expressions for oxidation rate and kinetics of oxygen consumption at different mechanisms of inhibitor action are given in Table 1.1.

1.3. Kinetics of inhibited autoxidation of hydrocarbons

An important distinction of autoxidation of a hydrocarbon is that its product (hydroperoxide) is an initiator which causes a progressively increasing initiation rate in the course of the reaction. The rate of acceleration depends in turn on the rate of chain oxidation; i.e., there is a kind of a positive feedback between the autoinitiation and autoxidation reactions. A similar feedback exists in inhibited oxidation of other organic compounds, too.

First, the more effectively the inhibitor terminates the chains, the slower is its rate of consumption and the longer is the effective inhibition period, τ, whereas in an initiated chain reaction, with v_i = const, it is independent of InH effectiveness. Second, autoxidation of RH can be inhibited not only with compounds that act to terminate the chains but also with compounds that decompose ROOH.[1,3] Unless it involves buildup of free radicals, such decomposition will hamper accumulation of ROOH and thereby inhibit the autoinitiation. Good peroxide decomposers include sulfur and phosphorus compounds and various metal complexes, e.g., thiophosphates and thiocarbamates of zinc, nickel, and other metals.[12,13] Third, critical phenomena are often observed in inhibited autoxidation experiments, which must be attributed to the above-mentioned feedback effect.[15]

Since ROOH is decomposed during autoxidation, the oxidation may follow either of two regimes, a nonsteady-state or a quasi-steady-state one with respect to ROOH. Under the non-steady-state process the ROOH is stable and almost no decomposition is perceptible during the induction period; that is its decomposition rate constant $k_3 < \tau^{-1}$. Obviously, such a regime arises due to specific conditions of inhibited oxidation, which will depend on the structure and reactivity of RH, ROOH and InH. Since oxidation of RH and consumption of InH are interrelated processes, the O_2 absorption rate may be quantitatively expressed in terms of the rate of consumption of InH in the system by using the following equations (v_{InH} is the rate of consumption of InH):

$$v_i = v_{i\,0} + k_3[\text{ROOH}],\ v_{InH} = v_i / f \qquad (1.13)$$

$$v = k_2[\text{RH}][RO_2^\bullet] + k_7[\text{InH}][RO_2^\bullet] \qquad (1.14)$$

For every possible mechanism of inhibited oxidation one may correlate [$RO_2^\bullet$] with [InH] and [ROOH], express the results mathematically and, after solving a set of two differential equations describing absorption of oxygen and consumption of InH, express the absorbed quantity of oxygen in terms of the consumed InH. For particular mechanism the correlation will have its special form. Table 1.2 contains formulas relating the degree of oxidation with consumed InH. The calculation was based on the following assumed conditions (i) $\Delta [O_2] = [ROOH]$, (ii) rate of initial chain generation $v_{i0} << k_3 [ROOH]$,

Table 1.2
Formulas for [ROOH] = F ([InH]) at inhibited hydrocarbon oxidation in nonstationary regime; $x = [InH] / [InH]_0$, $[InH] = [InH]_0$ at $t = 0$

Key steps	[ROOH]	a
2, 7	$[InH]_0(1 - x - a \ln x)$	$k_2 [RH] / k_7 [InH]_0$
2, −7, 8	$a ([InH]_0^{1/2} - [InH]^{1/2})$	$2 k_2 [RH] (f k_{-7} / k_3 k_7 k_8)^{1/2}$
2, 7, 8, 10	$a ([InH]_0^{1/2} - [InH]^{1/2})^{2/3}$	$(9 f k_2^2 k_{10} / k_3 k_7 k_8)^{1/3} [RH]$
2, 7, 11	$[InH]_0(b (1 - x) - a \ln x)$	$k_2 [RH] / k_7 [InH]_0$, $b = 2 k_2 k_{11} [RH] / f k_3 k_7$
2, 7, 12	$a [InH]_0(- b \ln x)$	$2 k_2 [RH] / k_7 [InH]_0$
		$b^{-1} = 1 + f/4(1+k_3 (1+a) / k_{12} [O_2])$
2, 7, 13	$a [InH]_0(- \ln x)$	$2 k_2 [RH] / k_7 [InH]_0$
2, 8, 14	$a\{ ([InH]_0^{1/2} - [InH]^{1/2})\}^{2/3}$	$(9 f k_2^2 k_{14} [RH]^2 / k_3 k_7 k_8)^{1/3}$
2, 9, 15	$[InH]_0(b (1 - x) - a \ln x)$	$k_2 [RH] / k_7 [InH]_0$
		$b^{-1} = 1 + [k_3 k_9 (1 + a)[InH]_0]^{1/2} / (k_{15} [O_2])$

and (iii) the mechanism remains unchanged through the induction period. Since the rate of inhibitor consumption $v_{InH} = v_i / f$ and v_i tends to increase during oxidation, the InH consumption kinetics is substantially nonlinear. During the early stages of oxidation $v_{InH} = v_{i0} / f$ but as more ROOH is accumulated v_{InH} increases and becomes maximum toward the end of the induction period. Calculation and experiment yield identical results.

At a sufficiently high temperature or in the presence of a ROOH decomposer, ROOH will rapidly dissociate and therefore the oxidation regime will quickly become quasi-steady as regards the ROOH concentration, with the decomposition rate equal to the rate of its formation. However, the ROOH concentration will tend to increase, since as more InH is consumed, the inhibition effect will decline and the ROOH formation rate will increase. A necessary condition for the quasi-steady process is the inequality $k_d \tau >> 1$, where k_d is the total rate constant of ROOH consumption by all possible routes, including dissociation to radicals, decomposition to molecular products, and decomposition under attack of free radicals. The change from a nonsteady to the quasi-steady condition is related to the induction period τ, which depends on the InH type and concentration. The transition from one inhibitor to another often manifests itself in transitions from one type of autoxidation process to another and various critical phenomena.[15]

What we mean by critical effects in inhibited autoxidation of RH is that under a certain critical InH concentration $[InH]_{cr}$ there takes place a sharp change in the τ vs. [InH] relationship; i.e., $d\tau/d[InH]$ for $[InH] > [InH]_{cr}$ is much greater than $d\tau/d[InH]$ for $[InH] < [InH]_{cr}$. Critical effects may arise when (i) inhibited oxidation proceeds via mechanisms, that include chain termination by reaction 7 (see Table 1.3), (ii) hydroperoxide is the main initiating agent so that $v_{io} << k_3[ROOH]$, and (iii) decomposition of ROOH is sufficiently rapid so the condition $k_3 >> \tau^{-1}$ for [InH] > $[InH]_{cr}$ is satisfied. As we have said earlier the critical phenomena are due to the feedback effect in inhibited oxidation and occur when both the formation and decomposition rates are similarly dependent upon [ROOH], for example, when they are directly proportional to [ROOH]. If the oxidation rate is proportional to $[ROOH]^n$, with $n < 1$, and the decomposition rate is proportional to [ROOH], the critical effects will never take place. Table 1.3 contains formulas for chain lengths ν, $[InH]_{cr}$ and the quasi-steady hydroperoxide concentration $[ROOH]_s$ for different inhibited oxidation mechanisms. The table 1.4 also contains formulas for the induction periods of inhibited autoxidation. The induction period was calculated as the time of InH decrease from $[InH]_0$ to $[InH]_{cr}$ (mechanisms 1, 3, 5) or to zero.

One may obtain a synergistic effect in inhibition of autoxidation of RH if two different InH, one of which acts to terminate the chains (ArOH, AmH) and the other of which decomposes hydroperoxides (sulfide, zinc carbamate, etc.,) are added to RH. In combination the two InH are operative for a longer period than either of them separately because one of them decelerates the buildup of ROOH by breaking the chains and the other, by decomposing ROOH, reduces v_i and slows down the consumption of the first InH. Antioxidant of complex function, containing groups reacting, with both $RO_2^{\bullet}$ and ROOH, may be expected to be highly effective.

Table 1.3
Formulas for kinetic parameters of hydrocarbon autoxidation as chain reaction in quasistationary regime: chain length ν, critical concentration of inhibitor $[InH]_{cr}$, and quasistationary concentration of hydroperoxide $[ROOH]_r$. The following symbols are used: $\beta = k_3/k_d$ and v_{i0} is rate of free radical generation on reaction of RH with oxygen

Key steps	ν	$[InH]_{cr}$	$[ROOH]_s$
2, 7	$k_2[RH]/fk_7[InH]$	$\beta k_2[RH]/fk_7$	$\beta \nu v_{i0}/k_3(1-\beta\nu)$
2, 8, 10	β^{-1}	—	$\beta^2 k_2 k_{10}[RH]^3/fk_3k_7k_8[InH]$
2, −7, 8	$k_2 k_{-7}^{1/2}[RH](fk_3k_7k_8[InH])^{1/2}$	$\beta^2 k_2^2 k_{-7}[RH]^2/fk_3k_7k_8$	$v_{i0}/fk_3\{([InH]/[InH]_{cr})-1\}$
2, 7, 11	$k_2[RH]/fk_7[InH]$	$\beta k_2[RH]/fk_7$	$\beta\nu v_{i0}/(1-\beta\nu)(k_3+k_{11}[InH])$
2, 7, 12	$k_2[RH]/fk_7[InH]$	$\beta k_2[RH]/fk_7$	$\beta\nu(v_{i0}+k_{12}[O_2][InH])/$ $/k_3(1-\beta\nu)$
2, 7, 13	$k_2[RH]/fk_7[InH]$	$\beta k_2[RH]/fk_7$	$2\beta\nu v_{i0}/k_3(1-2\beta\nu)$
2, 8, 14	β^{-1}	—	$\beta^2 k_2^2 k_{14}[RH]^2/fk_3k_7k_8[InH]$
2, 9, 15	$k_{15}[O_2](2k_9v_{i0})^{-1/2}(2+k_2k_7^{-1}[RH][InH]^{-1})$	$k_7(2k_9v_{i0})^{1/2}(\beta k_2k_{15}[RH])^{-1}$ $-2k_2^{-1}k_7[O_2][RH]^{-1}$	$\beta\nu v_{i0}/k_3(1-\beta\nu)$

Table 1.4
Formulas for induction period τ of inhibited oxidation of hydrocarbons in quasistationary regime. Symbols are the following: $\tau_0 = f\,[\mathrm{InH}]_0\, v_{i\,0}^{-1}$, $\beta = k_3 / k_d$, $v_{i\,0}$ is the rate of free radical generation on reaction of RH with oxygen

Key steps	τ/τ_0	x
2, 7, 8	$1 - x^{-1}\,(1 + \ln x)$	$f k_7\,[\mathrm{InH}]_0 / \beta k_2\,[\mathrm{RH}]$
2, 8,10	$1 - x^{-1}\,\ln (1 + x)$	$f k_7 k_8 [\mathrm{InH}]_0 v_{i\,0} / \beta^2 k_2^2 k_{10} [\mathrm{RH}]^3$
2, 7, 11	$[\mathrm{InH}]_0/[\mathrm{InH}]_{cr} - a \ln x$	$(k_7 k_{11} [\mathrm{InH}]_{cr} + b)[\mathrm{InH}]_0/(k_7 k_{11} [\mathrm{InH}]_0 + b)[\mathrm{InH}]_{cr}$;
		$a = \beta k_3\,[\mathrm{RH}][(f[\mathrm{InH}]_0)(k_2^{-1} k_3 k_7 + \beta k_{11}\,[\mathrm{RH}])^{-1}$;
		$b = k_3\, k_7 + \beta k_2\, k_{11}\,[\mathrm{RH}]$
2, 7, 12	$cx^{-1}\,(2 - b/a)\,\{\arctan [\,c(2ax + b)] -$	$f k_7\,[\mathrm{InH}]_0 / \beta k_2\,[\mathrm{RH}]$,
	$\arctan [\,c\,(2a + b)]\}$	$a = f k_7 k_{12}\,[\mathrm{O_2}][\mathrm{InH}]_{cr}^2$, $b = v_{i\,0} k_7\,[\mathrm{InH}]_{cr}$,
		$c^{-1} = b\,[4\,(f - 1)\,k_{12}\,[\mathrm{O_2}][\mathrm{InH}]_{cr}\,v_{i0}^{-1} - 1]^{1/2}$
2, −7, 8	$1 - x^{-1}\,\{1 + \ln (2 - 2x^{1/2} + x) + \arctan(x^{1/2} - 1)$	$f k_3\, k_7\, k_8\,[\mathrm{InH}]_0 / \beta^2 k_2^2 k_{-7}\,[\mathrm{RH}]^2$
2, 8, 14	$1 - x^{-1}\,\ln (1 + x)$	$f k_7\, k_8\,[\mathrm{InH}]_0 v_{i\,0} / \beta^2\, k_2^2\, k_{14}\,[\mathrm{RH}]^2$

1.4. Mechanisms of cyclic chain termination

As it terminates the chains, InH is consumed in the course of the oxidation process. If chain termination by reactions 7 and 8 (see scheme) dominate in the system and the products of reactions 8 are nonactive toward chain termination or initiation so ,$f = 2$. If $\mathrm{In}^\bullet$ decays by reaction 9, so $f = 1$. If reaction 8 products take part in chain termination, then it may be that $f > 2$, but as experience shows, for oxidizing hydrocarbons $f < 4$ for hydroquinones, amines, and diamines of the *p*-phenylenediamine type.[16] At the same time there are inhibitors that in a given system cause catalyzed termination of chains and in this case $f >> 2$.[3,17,18] Multiple chain termination was observed in oxidizing cyclohexanol with α-naphthylamine.[19] This was shown to be typical of a number of aromatic amines in primary and secondary alcohol oxidation.[20] The range of compounds in oxidation of which the InH becomes involved in multiple chain terminations is rather broad and includes also cyclohexadiene,[21] primary, secondary, and tertiary aliphatic amines,[22,23] cyclohexanone containing hydrogen peroxide,[24] and 1,2-disubstituted ethylenes.[25] For these compounds $f >> 2$ with AmH, nitroxyl radicals, certain ArOH, and quinones used as InH. The high nonstoichiometric values of f are due to the catalytic mechanism of chain termination occurring in such systems. The mechanism itself is due to the dual function of hydroxyperoxyl, aminoperoxyl, and hydroperoxyl radicals, which may either oxidize or reduce.

The pure catalytic chain termination occurs on copper ions in oxidizing cyclohexanol.[26]

$$>C(OH)OO^{\bullet} + Cu^{+} \rightarrow >C(OH)OO^{-} + Cu^{2+} \quad (16)$$

$$>C(OH)OO^{\bullet} + Cu^{2+} \rightarrow >C{=}O + O_2 + Cu^{+} + H^{+} \quad (17)$$

The value of rate constants of hydroxyperoxyl radical reactions with metallic ions and complexes are summarized in Table 1.5.

Table 1.5
Rate constants of peroxyl radical reaction with metal complexes

Metal complex	Oxidizing substance	T/ K	k/ $l\ mol^{-1}s^{-1}$	Ref.
Manganese (II) stearate	*cyclo*-[CH=CHCH=CH(CH_2)$_2$]	348	1.9×10^6	19
Manganese (II) stearate	*cyclo*-$C_6H_{11}OH$	348	2.4×10^6	59
Manganese (II) acetate	$(C_4H_9)_2NH$	348	2.5×10^6	60
Manganese (II) stearate	$(C_4H_9)_2NH$	348	3.5×10^6	24
Manganese (II) acetate	*cyclo*-$C_6H_{11}NH_2$	348	7.3×10^7	60
Manganese (II) stearate	*cyclo*-$C_6H_{11}NH_2$	348	1.6×10^8	24
Bis-(acetylacetonate)manganese (II)	*cyclo*-$C_6H_{11}NH_2$	363	9.8×10^7	61
N,N'-Ethylene-bis-(salicylideneiminato)manganese (II)	*cyclo*-$C_6H_{11}NH_2$	348	3.1×10^8	61
Manganese (II) acetate	$C_6H_5CH_2NH_2$	338	2.8×10^8	60
Manganese (II) acetate	$CH_2{=}C(CH_3)C(O)OC_2H_4N(CH_3)_2$	323	1.2×10^7	62
Manganese (II) acetate	$C_3H_7OCOCH_2CH_2N(CH_3)_2$	323	1.5×10^7	62
Bis-(dimethylglioximato)-bis-pyridineferrous	*cyclo*-[CH=CHCH=CH(CH_2)$_2$]	348	2.3×10^3	63
Bis-(dimethylglioximato)-bis-pyridineferrous	$(CH_3)_2CHOH$	344	1.0×10^3	64
Ferric stearate	*cyclo*-$C_6H_{11}OH$	348	4.8×10^3	59
Ferric stearate	*cyclo*-$C_6H_{11}NH_2$	348	1.2×10^6	24
Tris-(acetylacetonate)ferric	*cyclo*-$C_6H_{11}NH_2$	348	1.0×10^4	60
Bis-(dimethylglioximato)ammineiodidecobalt (II)	*cyclo*-[CH=CHCH=CH(CH_2)$_2$]	348	2.3×10^4	63
Bis-(dimethylglioximato)amminechloridecobalt (II)	*cyclo*-[CH=CHCH=CH(CH_2)$_2$]	348	8.4×10^2	63
Tris-(dimethylglioximato)cobalt (III)	$(CH_3)_2CHOH$	344	2.9×10^2	64
Bis-(dimethylglioximato)ammineiodidecobalt (II)	$(CH_3)_2CHOH$	344	3.1×10^2	64
Bis-(dimethylglioximato)pyridineiodidecobalt (II)	$(CH_3)_2CHOH$	344	7.0×10^4	64

Metal complex	Oxidizing substance	T/ K	k/ l mol^{-1}s^{-1}	Ref.
Bis-(dimethylglioximato)amminechloridecobalt (II)	$(CH_3)_2CHOH$	344	5.0×10^3	64
Bis-(dimethylglioximate)-bis-pyridinecobalt (II)	$(CH_3)_2CHOH$	344	1.2×10^4	64
Cobalt (II) stearate	*cyclo*-$C_6H_{11}OH$	348	1.0×10^5	59
Cobalt (II) chloride	*cyclo*-$C_6H_{11}OH$	348	6.2×10^4	19
Cobalt (II) acetate	*cyclo*-$C_6H_{11}OH$	348	1.3×10^5	19
Cobalt (II) stearate	*cyclo*-$C_6H_{11}OH$	348	3.1×10^4	19
Cobalt (II) cyclohexylcarboxilate	*cyclo*-$C_6H_{11}OH$	348	8.9×10^4	19
Bis-(acetylacetonato)cobalt (II)	*cyclo*-$C_6H_{11}OH$	348	6.4×10^4	60
N,N′-Ethylene-bis-(salicylideneiminato)cobalt (II)	*cyclo*-$C_6H_{11}OH$	348	3.3×10^5	61
Porfirine cobalt (II)	*cyclo*-$C_6H_{11}OH$	348	3.6×10^4	61
Bis-(salicylate) nickel	$C_8H_{17}OH$	363	1.6×10^4	61
Bis-(salicylate) nickel	*cyclo*-$C_6H_{11}NH_2$	353	2.8×10^4	61
N,N′-Ethylene-bis-(salicylideneiminato)nickel	$C_8H_{17}OH$	363	7.9×10^3	61
N,N′-Ethylene-bis-(salicylideneiminato)nickel	*cyclo*-$C_6H_{11}NH_2$	353	1.5×10^4	61
N,N′-Ethylene-bis-(salicylideneiminato)nickel	$C_8H_{17}OH$	363	3.5×10^4	61
N,N′-Ethylene-bis-(salicylideneiminato)nickel	*cyclo*-$C_6H_{11}NH_2$	353	6.6×10^4	61
N,N′-Ethylene-bis-(N-*p*-toluidinyl)salicylideneiminato)nickel	$C_8H_{17}OH$	363	8.0×10^4	61
N,N′-Ethylene-bis-(N-*p*-toluidinyl)salicylideneiminato)nickel	*cyclo*-$C_6H_{11}NH_2$	353	1.7×10^5	61
Copper stearate	*cyclo*-[CH=CHCH=CH$(CH_2)_2$]	348	1.9×10^6	19
Bis-(dimethylglioximato)copper	*cyclo*-[CH=CHCH=CH$(CH_2)_2$]	348	1.3×10^5	63
Bis-(diphenylglioximato)copper	*cyclo*-[CH=CHCH=CH$(CH_2)_2$]	348	1.5×10^5	63
Bis-(salicylate)copper	*cyclo*-[CH=CHCH=CH$(CH_2)_2$]	348	1.7×10^5	63
Bis-(dimethylglioximato)copper	$(CH_3)_2CHOH$	344	5.0×10^5	64
Bis-(salicylate)copper	$(CH_3)_2CHOH$	344	1.2×10^5	64
Bis-(diphenylglioximato)copper	$(CH_3)_2CHOH$	344	5.9×10^4	64
Bis-(dimethylglioximato)ammineiodidecopper	$(CH_3)_2CHOH$	344	3.7×10^4	64
Copper sulfate	*cyclo*-$C_6H_{11}OH$	348	3.2×10^6	23

Metal complex	Oxidizing substance	T/ K	k/ l mol^{-1}s^{-1}	Ref.
Copper stearate	*cyclo*-$C_6H_{11}OH$	348	1.1×10^6	59
Copper stearate	$(C_4H_9)_2NH$	348	4.1×10^6	24
Copper acetate	$C_6H_5CH_2NH_2$	338	1.5×10^8	60
Copper stearate	*cyclo*-$C_6H_{11}NH_2$	348	1.0×10^7	24
Bis-(acetylacetonato)copper	*cyclo*-$C_6H_{11}NH_2$	348	7.4×10^5	60
Porfirine copper	*cyclo*-$C_6H_{11}NH_2$	348	7.3×10^5	61
Copper acetate	$C_3H_7COOC_2H_4N(CH_3)_2$	323	1.3×10^7	62
Copper acetate	$C_4H_9N(CH_3)_2$	323	1.0×10^7	62
Copper stearate	*cyclo*-$C_6H_{11}OH$	348	4.0×10^4	59
Copper stearate	*cyclo*-$C_6H_{11}NH_2$	348	7.2×10^5	61

1,4-Benzoquinone (Q) multiply terminates the chains of oxidizing isopropyl alcohol[35]

$$Q + (CH_3)_2C(OH)OO^{\bullet} \rightarrow QH^{\bullet} + O_2 + (CH_3)_2C{=}O \quad (18)$$

$$QH^{\bullet} + {}^{\bullet}OOC(OH)(CH_3)_2 \rightarrow Q + HOOC(OH)(CH_3)_2 \quad (19)$$

Nitroxyl radical is reduced to the corresponding hydroxylamine, and the latter is oxidized back to the nitroxyl radical with the peroxyl radicals of cyclohexylamine.[35]

$$>NO^{\bullet} + H_2N({}^{\bullet}OO)C< \rightarrow >NOH + HN{=}C< + O_2 \quad (20)$$

$$>NOH + ROO^{\bullet} \rightarrow >NO^{\bullet} + ROOH \quad (21)$$

It is remarkable that oxidation-reduction activity is also exhibited by peroxyls that have next to their peroxyl group a heteroatom with a lone electron pair or a double bond. Such peroxyls probably react with the inhibitor radical, quinone, or variable-valence metal ion by way of electron transfer, e.g.

$$>C(OO^{\bullet})NHR + Cu^{2+} \rightarrow >C(OO^{\bullet})NHR^{+\bullet} + Cu^{1+} \quad (22)$$

followed by rapid elimination of the proton and oxygen molecule

$$>C(OO^{\bullet})N^{\bullet+}HR \rightarrow >C{=}NR + O_2 + H^+ \quad (23)$$

All the peroxyls we have mentioned decompose, generating $HO_2^{\bullet}$

$$>C(OH)OO^{\bullet} \rightarrow >C{=}O + HO_2^{\bullet} \quad (24)$$

$$>C(NH_2)OO^{\bullet} \rightarrow >C{=}NH + HO_2^{\bullet} \quad (25)$$

Thus, generated in the system and participating in catalyzed chain termination (if the corresponding InH has been added) are both hydroxyperoxides and hydroperoxides. Depending on the type of compound and oxidation conditions (temperature, degree of conversion), either $HO_2^•$ or $>C(OH)OO^•$ [$>C(NH_2)OO^•$] radicals may preferentially take part in chain termination. Things are just the same with chain termination in oxidizing amines. Aliphatic and alkyl aromatic peroxyl radicals do not take part in reactions of this sort because they have no reducing activity. Since the hydroxyperoxyl (hydroperoxyl, aminoperoxyl) radical behaves as both an oxidant and a reductant, it will react with $In^•$ by two parallel routes. In the case of AmH the following three mechanisms of chain termination appear to be probable.[37]

(i)
$$Ar_2NH + HO(^•OO)CR_2 \rightarrow Ar_2N^• + R_2C(OH)OOH \quad (26)$$
$$Ar_2N^• + HO(^•OO)CR_2 \rightarrow Ar_2NH + O_2 + R_2C{=}O \quad (27)$$

(ii)
$$Ar_2N^• + HO(^•OO)CR_2 \rightarrow Ar_2NO^• + HO(^•O)CR_2 \quad (28)$$
$$Ar_2NO^• + HO(^•OO)CR_2 \rightarrow Ar_2NOH + O_2 + R_2C{=}O \quad (29)$$
$$Ar_2NOH + HO(^•OO)CR_2 \rightarrow Ar_2NO^• + R_2C(OOH)OH \quad (30)$$

(iii)
$$C_6H_5ArN^• + HO(^•OO)CR_2 \rightarrow ArN{=}C_6H_4O + R_2CO + H_2O \quad (31)$$
$$OC_6H_4{=}NAr + HO(^•OO)CR_2 \rightarrow ArNHC_6H_4O^• + O_2 + R_2C{=}O \quad (32)$$
$$ArNHC_6H_4O^• + HO(^•OO)CR_2 \rightarrow ArNC_6H_4O + R_2C(OOH)OH \quad (33)$$

Multiple involvement in chain termination reactions has also been observed for the stable nitroxyl radicals 2,2,6,6-tetramethylpiperidin-N-oxyl and its derivatives.[38] Nitroxyl radicals ($>NO^•$) break the chains in oxidizing hydrocarbons and polymers by reacting with alkyl radicals.

$$>NO^• + R^• \rightarrow >NOR \quad (34)$$

If we accept this mechanism, the rate of inhibited polymer oxidation must be $v = k_1 [O_2] v_i / k_{34} [>NO^•]$. The k_{34} / k_1 ratio for 2,2,6,6-tetramethyl-4-benzoylpiperidin-N-oxyl equals 0.09 in polypropylene at 387 K and 0.027 in polyethylene at 365 K.[39] Nitroxyl radicals react slower than oxygen with alkyl radicals, and their high inhibiting effect in polymers is due to the relatively low dissolved oxygen concentration in polymers. The multiple involvement of $>NO^•$ in chain termination manifests itself in the fact that the rate of $>NO^•$ consumption in a polymer is much smaller than the rate of initiation under conditions where all the chains are terminated via reaction alkyl radical with nitroxyl.[38,39] Under the same conditions, but in the absence of oxygen, the nitroxyl radical is consumed at a rate equal to that of initiation. In an initiator-containing polymer all the nitroxyl radicals will be consumed within a time equal to $[>NO^•]_0 / v_i$ (as monitored by ESR). However, $>NO^•$ reappears in the system to which oxygen has been admitted, under attack of $RO_2^•$.[38,39] Regeneration has been proposed to be due to the following reaction[65]

$$RO_2^• + HCCON< \rightarrow ROOH + >C{=}C< + >NO^• \quad (35)$$

This supposition is consistent with the following data. The $>NO^•$ are formed from the products of reaction between the alkyl macroradical with $>NO^•$ only in the presence of O_2 and initiator, that is, under attack of $RO_2^•$. The product cannot be the corresponding hydroxylamine since it could not be extracted from the polymer with a solvent.

The following reaction has been demonstrated experimentally.[40]

$$[(CH_3)_3C]_2NOC(CH_3)_3 + {}^•O_2C(CH_3)_3 \rightarrow [(CH_3)_3C]_2NO^• + CH_2{=}C(CH_3)_2 + HO_2C(CH_3)_3 \quad (36)$$

The k_{36} value is rather high; it is 44 l mol^{-1} s^{-1} at 403 K in *tert*-butylbenzene. The scission of the weaker secondary C—H bond should have been faster.

A different regeneration mechanism was proposed in ref. 40, where it was noted that hydroxamic ether was thermally unstable and dissociated at the O—C bond, which was followed by cage disproportionation of radicals

$$R_2NOC(CH_3)_3 \rightarrow [R_2NO^\bullet + HCH_2C^\bullet(CH_3)_2] \rightarrow R_2NOH + CH_2{=}C(CH_3)_2 \quad (37)$$

By reacting with $RO_2^\bullet$ hydroxylamine was converted to $>NO^\bullet$. Below 400 K decomposition of hydroxamic ethers is slow (at 403 K it is decomposed with rate constant [40] $k_{37} = 5.7 \times 10^{-1}$ s^{-1}) and cannot be responsible for the experimentally observed regeneration rates. Besides, under the conditions that had been used in the polypropylene experiments[39] this mechanism runs contrary to some of the above factors. Yet at a higher temperature when hydroxamic ether decomposes faster, this mechanism might possibly become effective.

An increase of the stoichiometric coefficient f has also been observed for ArOH in polymers as the partial O_2 pressure was reduced. For example, $f = 1$ and 3.3 for α-naphthol in oxidizing polypropylene at 388 K at $Po_2 = 100$ and 0 kPa, respectively.[41] The reason is that the reduction of $[O_2]$ increases the fraction of $ArO^\bullet$ reacting with alkyl radicals, which not only recombine but also disproportionate

$$ArO^\bullet + R_2CHCR_2^\bullet \rightarrow ArOH + R_2C{=}CR_2 \quad (38)$$

Thus the mechanisms by which the inhibitor radicals are regenerated in chain termination are quite varied.

The acid catalyzed cyclic chain termination by nitroxyl radicals were found in oxidizing hydrocarbons recently.[42,43] Inhibiting system includes nitroxyl radical, hydrogen peroxide and organic acid. Effective chain termination provoke only triple system and binary systems ($AmO^\bullet + H_2O_2$, $AmO^\bullet$ + acid (HA), H_2O_2 + acid) have very weak inhibiting effect on ethylbenzene oxidation. Hydrogen peroxide is consumed during induction period and nitroxyl radical and acid practically are not decomposed during oxidation. The following mechanism was proposed.

$$AmO^\bullet + HA \Longleftrightarrow AmOH^{\bullet+} + A^- \quad (39)$$
$$RO_2^\bullet + AmOH^{\bullet+}, A^- \rightarrow ROOH + AmO^+, A^- \quad (40)$$
$$AmO, A^- + H_2O_2 \rightarrow AmOH + O_2 + HA \quad (41)$$
$$RO_2^\bullet + AmOH \rightarrow ROOH + AmO^\bullet \quad (42)$$

Protonization of nitroxyl radical transforms it into the form reactive toward peroxyl radical and formed nitronium ion is reduced fast by hydrogen peroxide. These two reactions together with fast reaction of formed hydroxyamine with $RO_2^\bullet$ give rise the cycle of effective chain termination. The same mechanism apparently takes place when the triple system: alcohol + nitroxyl radical + acid is introduced into oxidizing hydrocarbon.[42] Very close to this mechanism is that of another triple inhibiting system: iminoquinone (Q) + hydrogen peroxide + acid.[44] The following mechanism is proposed for action of this system.

$$Q + HA \Longleftrightarrow QH^+ + A^- \quad (43)$$
$$RO_2^\bullet + QH^+, A^- \rightarrow ROOH + Q^{\bullet+}, A^- \quad (44)$$
$$Q^{\bullet+}, A^- + H_2O_2 \rightarrow {}^\bullet QH + O_2 + HA \quad (45)$$
$$RO_2^\bullet + {}^\bullet QH \rightarrow ROOH + Q \quad (46)$$

1.5 The parabolic transition state model as semiempirical method of evaluation of activation energies of free radical reactions with hydrogen atom abstraction

Among different empirical and semiempirical methods of evaluation of rate constants the parabolic transition state model of free radical reaction of atom abstraction[45] is rather simple, convenient and gives reliable results. The main theses of this conception are the following.[45,46] In a reaction of the type

$$R^{\bullet}_f + HR_i \rightarrow R_fH + R_i^{\bullet} \tag{47}$$

the R_i—H bond is being broken and the R_f—H bond is being formed. According to the theory of absolute rates, the reaction may be treated as a translation of the hydrogen atom along the reaction coordinate from an initial position at $x = 0$ with potential energy U_i (0) = 0 to its final position at $x = r_e$ and U_i $(r_e) = \Delta H_{ei}$, where ΔH_{ei} is the reaction enthalpy, with zero energies taken into account, so

$$\Delta H_{ei} = D_i - D_f + 0.5hL(v_i - v_f) \tag{1.15}$$

where D_i and D_f are the dissociation energies of the R_i—H and R_f—H bonds, v_i and v_f are their vibration frequencies, h is Planck's constant and L is Avogadro's number. Let us consider the vibration of atoms along the R_i—H and R_i—H bonds to be harmonic, so that $U_i^{1/2} = b_i x$ and $U_f^{1/2} = b_f (r_e - x)^{1/2}$, $b_f, = \pi v_i (2\mu_i)^{1/2}$, and $b_f = \pi v_f (2\mu_f)^{1/2}$ where μ_i and μ_f are the reduced masses of the atoms. The transition state is proposed to be represented by the point of intersection of two undisturbed potential curves at $x = r^{\#}$, when $E_{ei} = U_i (r^{\#}) = U_i (r_e - r^{\#}) - \Delta H_{ei}$.. The activation energy E_{ei} is related to the observed E_i by the equation.

$$E_e = E + 0.5(hv_i L - RT) \tag{1.16}$$

Activation energy E_i in its turn may be calculated on the value of experimentally found rate constant k using Arrhenius equation

$$E = RT \ln(A/k) \tag{1.17}$$

where A is preexponential factor of a given group of free radical reaction known from experimental measurements. The important characteristic of the transition state is the distance r_e which may be estimated in the form br_e using eqn. (1), where $\alpha = b_i/b_f = v_I v_f^{-1} (\mu_i / \mu_f)^{1/2}$

$$b_i r_e = \alpha (E_{ei} - \Delta H_{ei})^{1/2} + E_{ei}^{1/2} \tag{1.18}$$

The activation energy of the thermoneutral reaction in the given series may be calculated, using eqn.

$$E_{eo} = (b_i r_e)^2 (1 + \alpha)^{-2} \tag{1.19}$$

The parameter $b_i r_e$ allows us to calculate E_{ei} for any reaction using eqn. (1.20).

$$E_{ei} = (b_i r_e)^2 (1 - \alpha^2)^{-2} \{1 - \alpha[1 - (1 - \alpha^2)(b_i r_e)^{-2} \Delta H_e]^{1/2}\}^2 \tag{1.20}$$

Eqn. 1.20 takes it simplest form when $\Delta H_e << (b_i\, r_e)^2\, (1-\alpha^2)^{-1}$

$$E_{ei}^{\,1/2} = b_i\, r_e\, (1+\alpha)^{-1} + \alpha\, \Delta H_{ei}\, (2b_i\, r_e)^{-1} \qquad (1.21)$$

The above equations are valid for calculating activation energy of any free radical reaction with ΔH_e that is limited by minimum and maximum values, namely at $\alpha = 1$.

$$\Delta H_{e\,min} = b_i r_e\, (2hL\nu_i)^{1/2} - (b_i r_e)^2 \qquad (1.22)$$

$$\Delta H_{e\,max} = (b_i r_e)^2 - b_i r_e (2hL\nu_i)^{1/2} \qquad (1.23)$$

At $\Delta H_e < \Delta H_{e\,min}$ activation energy $E = 0$, and at $\Delta H_e > \Delta H_{e\,max}$ activation energy $E = \Delta H$. These equations were used to calculate rate constants of hydrogen atom abstraction from RH, ArOH, AmH etc. by different radicals, that are generated in oxidizing hydrocarbons in the presence of inhibitors. The values of parameters br_e, , $E_{e\,0}$ and r_e are given in the Table 1.6. All these values were calculated from experimental data (rate constants) in hydrocarbon solutions.

Table 1.6
Kinetic parameters of free radical reactions with hydrogen atom abstraction in the parabolic model of transition state

Reaction	br_e/ (kJ mol^{-1})$^{1/2}$	α	A/ l mol^{-1} s^{-1}	E_{e0}/ kJ mol^{-1}	$r_e \times 10^{11}$/ m	$b_i/(b_i + b_f)$
$RO_2^\bullet + R_1H$	14.23	0.814	1.0×10^8	61.5	3.802	0.449
$RO_2^\bullet + R_2H$	15.68	0.814	1.0×10^7	74.7	4.189	0.449
$RO_2^\bullet + R_3H$	14.74	0.814	1.0×10^7	66.0	3.938	0.449
$RO_2^\bullet$ + ROOH	13.13	1.00	1.0×10^8	43.1	2.854	0.500
$RO_2^\bullet + Ar_1OH$	13.46	1.00	3.2×10^7	45.3	2.885	0.500
$RO_2^\bullet + Ar_2OH$	14.40	1.00	3.2×10^7	51.8	3.087	0.500
$RO_2^\bullet$ + AmH	12.12	0.940	1.0×10^8	39.0	2.802	0.485
$RO_2^\bullet$ + AmOH	13.50	1.00	3.2×10^7	45.6	3.318	0.500
$RO_2^\bullet$ + ArSH	10.39	0.658	3.2×10^7	39.3	3.434	0.420
$Ar_1O^\bullet + R_3H$	15.53	0.802	1.0×10^8	74.3	4.162	0.445
$Ar_1O^\bullet + Ar_2OH$	13.31	1.00	1.0×10^8	51.6	3.080	0.500
$Ar_1O^\bullet$ + AmH	10.13	0.927	1.0×10^8	27.6	2.343	0.481
$Ar_1O^\bullet$ + AmOH	12.93	1.00	1.0×10^8	41.8	2.772	0.500
$Ar_1O^\bullet$ + ROOH	13.46	1.00	1.0×10^8	45.3	2.926	0.500
$Ar_1O^\bullet$ + ArSH	10.48	0.649	1.0×10^8	40.5	3.463	0.606

Reaction	br_e / $(kJ\ mol^{-1})^{1/2}$	α	A / $l\ mol^{-1}\ s^{-1}$	E_{eo} / $kJ\ mol^{-1}$	$r_e \times 10^{11}$ / m	$b_i/(b_i + b_f)$
$Ar_2O^{\bullet} + R_3H$	18.06	0.802	1.0×10^8	100.4	4.823	0.445
$Ar_2O^{\bullet} + Ar_1OH$	13.31	1.00	1.0×10^8	44.3	2.853	0.500
$Ar_2O^{\bullet} + Ar_2OH$	14.37	1.00	1.0×10^8	51.6	3.080	0.500
$Ar_2O^{\bullet}$ + AmOH	14.54	1.00	1.0×10^8	52.9	3.117	0.500
$Ar_2O^{\bullet}$ + ROOH	14.40	1.00	3.2×10^7	51.8	3.130	0.500
$Ar_2O^{\bullet}$ + AmH	11.26	0.927	1.0×10^8	34.2	2.605	0.481
$Am^{\bullet}$ + AmH	11.63	1.00	1.0×10^8	33.8	4.233	0.500
$Am^{\bullet}$ + AmOH	11.97	1.079	1.0×10^7	33.2	2.566	0.519
$Am^{\bullet}$ + ROOH	12.89	1.064	1.0×10^8	39.0	2.802	0.515
$Am^{\bullet} + R_3H$	16.87	0.866	1.0×10^8	81.7	4.507	0.464
$AmO^{\bullet} + Ar_1OH$	12.93	1.00	1.0×10^8	41.8	2.772	0.500
$AmO^{\bullet} + Ar_2OH$	14.54	1.00	1.0×10^8	52.9	3.117	0.500
$AmO^{\bullet}$ + AmH	11.10	0.927	1.0×10^8	33.2	2.568	0.481
$AmO^{\bullet}$ + AmOH	12.61	1.00	3.2×10^7	39.8	2.703	0.500
$AmO^{\bullet}$ + ROOH	13.50	1.00	1.0×10^8	45.6	2.894	0.500
$AmO^{\bullet} + R_1H$	13.72	0.802	1.0×10^9	58.0	3.665	0.445
$AmO^{\bullet} + R_2H$	15.66	0.802	1.0×10^8	75.5	4.184	0.445
$AmO^{\bullet} + R_3H$	14.42	0.802	1.0×10^8	64.0	3.852	0.445
$AmO^{\bullet}$ + ArSH	11.58	0.649	1.0×10^8	49.3	3.827	0.394
$ArS^{\bullet}$ + AmOH	17.84	1.541	1.0×10^8	49.3	3.827	0.606
$ArS^{\bullet} + Ar_1OH$	16.07	1.534	1.0×10^8	40.2	3.768	0.395
$ArS^{\bullet}$ + ROOH	15.79	1.520	3.2×10^7	39.3	3.433	0.603
$ArS^{\bullet} + R_3H$	13.74	1.238	1.0×10^8	37.7	3.668	0.553

REFERENCES

1 **Ingold, K. U.**, The retarding action of inhibitors on autoxidation of organic compounds in liquid phase, *Chem. Rev.*, 61, 563, 1961.
2 **Emanuel, N. M., Denisov, E. T., Maizus, Z. K.**, *Liquid Phase Oxidation of Hydrocarbons*, Plenum, New York, 1967, Chap. 7.
3 **Denisov, E. T., Mitskevich, N. I., Agabekov, V. E.**, *Liquid-Phase Oxidation of Oxygen-Containing Compounds*, Consultants Bureau, New York, 1977, Chap. 2.
4 **Mill, T., Hendry, D. G.**, In *Comprehensive Chemical Kinetics*, Elsevier, Amsterdam, vol. 16, 1980, Chap. 1.
5 **Emanuel, N. M., Zaikov, G. E., Maizus, Z. K.**, *Oxidation of Organic Compounds. Effect of Medium*, Pergamon, Oxford, 1984, Chap. 7.
6 **Kucher, R. V., Opeida, I. A.**, *Cooxidation of organic compounds in liquid-phase*, Naukova Dumka, Kiev, 1989, 208 (in Russian).
7 **Landberg, W. O., Ed.**, *Autooxidation and Antioxidants*, Interscience, New York, 1962, vol. 1.
8 **Scott, G.**, *Atmospheric Oxidation and Antioxidants*, Elsevier, Amsterdam, 1965, Chap. 4.
9 **Howard, J. A.**, Absolute rate constants for reactions of oxyl radicals, *Adv. Free Radical Chem.*, 4, 49, 1972.
10 **Denisov, E. T., Kovalev, G. I.**, *Oxidation and Stabilization of Jet Fuels*, Khimiya, Moscow, 1990, 270 (in Russian).
11 **Roginskii, V. A.**, *Phenolic antioxidants. Reactivity and effectiveness*, Nauka, Moscow, 1988, 248 (in Russian).
12 **Denisov, E. T.**, *Oxidation and degradation of carbonchain polymers*, Khimiya, Leningrad, 1990, 287 (in Russian).
13 **Scott, G., Ed.**, *Mechanism of polymer degradation and stabilization*, Elsevier Applied Science, London, 1990, 329.
14 **Denisov, E. T., Khudyakov, I. V.**, Mechanism of action and reactivity of the free radicals of inhibitors, *Chem. Rev.*, 87, 1313, 1987.
15 **Emanuel, N. M., Gagarina, A. B.**, Critical phenomena in chain reaction with degenerate branching, *Usp. Khim.*, 35, 619, 1966.
16 **Denisov, E. T.**, *Liquid-Phase Reaction Rate Constants*, Plenum, New York, 1974, Chap. 6.
17 **Denisov, E. T.**, Regeneration of inhibitors and negative catalysis in chain reactions of oxidation, *Kinet. Katal.*, 11, 312, 1970.
18 **Denisov, E. T.**, in *Developments in polymer stabilization-3*, Applied Science Publishers, London, 1979, Chap. 1.
19 **Denisov, E. T., Kharitonov, V. V.**, Peculiarities of retarding action of 1-naphthylamine in oxidizing cyclohexanole, *Izv. Akad. Nauk SSSR., Ser.Khim.*, 2222, 1963.
20 **Denisov, E. T., Scheredin, V. P.**, Synergistic action of alcohols on inhibiting activity of aromatic amines, *Izv. Akad. Nauk SSSR, Ser.Khim.*, 919, 1964.
21 **Vardanyan, R. L., Denisov, E. T.**, Regeneration of inhibitors in oxidizing 1,3-cyclohexadiene, *Izv. Akad. Nauk SSSR, Ser. Khim.*, 2818, 1971.
22 **Kovtun, G. A., Alexandrov, A. L.**, Oxidation of aliphatic amines by molecular oxygen in liquid phase. 1. Kinetic of oxidation of primary and secondary amines., *Izv.Akad. Nauk SSSR,Ser.Khim.*, 2208, 1973.
23 **Kovtun,G. A., Alexandrov, A. L.**, Oxidation of aliphatic amines by molecular oxygen in liquid phase . 4. Regeneration of inhibitors in oxidizing tertiary amines, *Izv. Akad. Nauk SSSR, Ser. Khim.*, 1274, 1974.
24 **Kharitonov, V. V., Denisov, E. T.**, The dual reactivity of hydroxyperoxyl radicals in reaction with aromatic amines, *Izv. Akad. Nauk SSSR, Ser. Khim.*, 2764, 1967.
25 **Sokolov, A. B., Nikanorov, A. A., Pliss, E. M., Denisov, E. T.**, Effect of multidipoles interaction in reactions with peroxyl radicals, *Izv. Akad. Nauk SSSR, Ser. Khim.*, 778, 1985.
26 **Alexandrov, A. L., Denisov, E. T.**, Negative catalysis by cupric ions in chain oxidation of cyclohexanole, *Izv. Akad. Nauk SSSR, Ser. Khim.*, 1652, 1969.
27 **Alexandrov, A. L., Solov'ev, G. I., Denisov, E. T.**, Negative catalysis by heavy metal stearates in chain oxidation of cyclohexanole, *Izv. Akad. Nauk. SSSR, Ser. Khim.*, 1527, 1972.
28 **Kovtun, G. A., Lukoianova, G. L., Berenblum, A. S., Moiseev, I. I.**, Antioxidative properties of nickel complexes of salicylaldoximes, *Dokl. Akad. Nauk. SSSR*, 231, 656, 1976.
29 **Kovtun, G. A., Alexandrov, A. L., Denisov, E. T.**, Oxidation of aliphatic amines by molecular oxygen. Kinetics of regeneration of inhibitors in oxidizing primary and secondary amines., *Izv. Akad.Nauk. SSSR, Ser. Khim.*, 2611, 1973.
30 **Kovtun, G. A.**, *Complexes of transition metals as catalysts of chain termination in oxidation*, Doct. Sci. (Chem.) Thesis Dissertation, Moscow, 1984, p. 14–24 (in Russian).
31 **Pliss, E. M., Alexandrov, A. L.**, Negative catalysis by heavy metals in oxidizing tertiary aliphatic amines., *Izv. Akad. Nauk. SSSR, Ser. Khim.*, 214, 1978.
32 **Zubareva, N. G., Denisov, E. T., Ablov, A. V.**, Negative catalysis by dioximines of transition metals in oxidizing 1,3-cyclohexadiene, *Kinet. Katal.*,14, 579, 1973.
33 **Zubareva, N. G., Denisov, E. T., Ablov, A. V.**, Negative catalysis by dioximines of transition metals in oxidizing isopropanole., *Kinet. Katal.*,14, 346, 1973.

34 **Kovtun, G. A.,** *Mechanism of oxidation of aliphatic amines and regeneration of antioxidants.* Cand. Sci. (Chem.) Thesis Dissertation, Inst. Chem. Phys., Chernogolovka, 1974, 164 (in Russian).

35 **Denisov, E. T.,** Mechanism of inhibiting action of quinone in chain oxidation of isopropanole, *Izv. Akad. Nauk SSSR, Ser. Khim.,*328, 1969.

36 **Kovtun, G. A., Golubev, V. A., Alexandrov, A. L.,** Kinetic study of regeneration of nitroxyl radicals in primary and secondary amines, *Izv. Akad.Nauk SSSR, Ser. Khim.,*793, 1974.

37 **Denisov, E. T., Goldenberg, V. I., Verba, L. G.,** Mechanism of cyclic chain termination and intermediates of aromatic amines in oxidizing isopropanole and ethylbenzene, *Izv. Akad. Nauk SSSR, Ser. Khim.*, 2217, 1988.

38 **Shilov, Yu. B., Battalova, R. M., Denisov, E. T.,** Regeneration of nitroxyl radicals in oxidizing polypropylene, *Dokl.Akad. Nauk SSSR,* 207, 388, 1972.

39 **Shilov,Yu. B., Denisov, E. T.,** Mechanism of inhibiting action of nitroxyl radicals in oxidizing polyethylene and polypropylene, *Vysokomol. soedin., Ser.A* , 16, 2313. 1974.

40 **Berger, H., Bolsman,T. A. B. M., Brouwer, D. M.,** in *Developments in polymer stabilization-6,* Ed. Scott G., Applied Science Publishers, London, 1983, Chap. 1.

41 **Zolotova, N. V., Denisov, E. T.,** Kinetic peculiarities of inhibiting action of phenols in polypropylene oxidation, *Vysokomol. soedin., Ser.B,* 18, 605, 1976.

42 **Goldenberg, V. I., Katkova, N. V., Denisov, E. T.,** Cyclic chain termination by nitroxyl radicals in oxidizing ethylbenzene in the presence of alcohols and acids, *Izv. Akad.Nauk SSSR, Ser. Khim.*, 287, 1988.

43 **Goldenberg, V. I., Denisov, E. T., Ermakova, N. A.,** Acid-catalyzed cyclic chain termination by nitroxyl radicals in hydrogen peroxide containing oxidizing ethylbenzene, *Izv. Akad.Nauk SSSR, Ser. Khim.*, 738, 1990.

44 **Goldenberg, V. I., Ermakova, N. A., Denisov, E. T.,** Acid-catalyzed cyclic chain termination by quinonemonoanilide in oxidizing hydrocarbons, *Izv. Akad. Nauk SSSR., Ser. Khim.* 79, 1995.

45 **Denisov, E. T.,** Nonlinear correlations in reactions of alkyl radicals with C—H bonds of organic compounds, *Kinet. Katal.*, 32, 461, 1991.

46 **Denisov, E. T.,** The parabolic transition state model and resultant nonlinear correlations for the kinetics of free radical reactions, *Mendeleev Commun.*, 2, 1, 1992.

47 **Denisov, E. T., Denisova, T. G.,** Kinetic parameters of reaction $RO_2^\bullet$ + RH on the principles of parabolic model of transition state, *Kinet. Katal.*, 34, 199, 1993.

48 **Denisov, E. T., Drozdova, T. I.,** Analysis of kinetic data of peroxyl radical reactions with phenols on the principals of parabolic model, *Kinet. Katal.*, 35, 176, 1994.

49 **Denisov, E. T.,** Triplet repulsion and electron affinity as factors determining activation energy of free radical abstraction reaction, *Kinet. Katal.*,35, 325, 1994.

50 **Denisov, E. T.,** Main factors determining reactivity of reagents in free radical abstraction reaction, *Kinet. Katal.*, 35, 671, 1994.

Chapter 2

RATE CONSTANTS OF ELEMENTARY STEPS OF CHAIN OXIDATION OF HYDROCARBONS

Table 2.1
Enthalpies, activation energies and rate constants of reaction $RO_2^{\bullet} + RH \rightarrow ROOH + R^{\bullet}$, E calculated by formulas 1.15–1.17, 1.21. The values of A, br_e and α, see Table 1.6

Hydrocarbon	$RO_2^{\bullet}$	ΔH/ kJ mol^{-1}	E/ kJ mol^{-1}	k (333 K)/ l mol^{-1} s^{-1}	k (400 K)/ l mol^{-1} s^{-1}
$CH_3(CH_2)_3CH_3$	$HO_2^{\bullet}$	34.3	60.3	0.21	8.0
$CH_3(CH_2)_3CH_3$	*sec*-$RO_2^{\bullet}$	37.8	62.2	0.11	4.5
$CH_3(CH_2)_3CH_3$	*tert*-$RO_2^{\bullet}$	44.7	65.8	2.3×10^{-2}	1.5
$(CH_3)_2CHCH_2CH_3$	HO_2.	23.2	54.8	0.25	7.0
$(CH_3)_2CHCH_2CH_3$	*sec*-$RO_2^{\bullet}$	26.7	56.5	0.14	4.2
$(CH_3)_2CHCH_2CH_3$	*tert*-$RO_2^{\bullet}$	33.6	60.0	3.8×10^{-2}	1.5
$C(CH_3)_4$	$HO_2^{\bullet}$	49.4	68.3	2.4×10^{-2}	1.4
$C(CH_3)_4$	*sec*-$RO_2^{\bullet}$	52.9	70.3	1.1×10^{-2}	0.8
$C(CH_3)_4$	*tert*-$RO_2^{\bullet}$	59.8	74.1	2.8×10^{-3}	0.3
cyclo-C_5H_{10}	$HO_2^{\bullet}$	26.4	56.4	1.42	4.3
cyclo-C_5H_{10}	*sec*-$RO_2^{\bullet}$	29.9	58.1	0.77	25.9
cyclo-C_5H_{10}	*tert*-$RO_2^{\bullet}$	36.8	61.6	0.22	9.0
cyclo-C_6H_{12}	$HO_2^{\bullet}$	30.6	58.5	0.80	27.5
cyclo-C_6H_{12}	*sec*-$RO_2^{\bullet}$	34.1	60.2	0.44	16.5
cyclo-C_6H_{12}	*tert*-$RO_2^{\bullet}$	41.0	63.8	0.18	5.6
cyclo-$C_6H_{11}CH_3$	$HO_2^{\bullet}$	20.3	53.4	1.22	1.3×10^{2}
cyclo-$C_6H_{11}CH_3$	*sec*-$RO_2^{\bullet}$	23.8	55.1	0.67	76.6
cyclo-$C_6H_{11}CH_3$	*tert*-$RO_2^{\bullet}$	30.7	58.5	0.25	27.5
$CH_2{=}CHCH_3$	$HO_2^{\bullet}$	–5.8	54.7	7.9×10^{-2}	2.2

Hydrocarbon	$RO_2^{\bullet}$	ΔH/ kJ mol^{-1}	E/ kJ mol^{-1}	k (333 K)/ l mol^{-1} s^{-1}	k (400 K)/ l mol^{-1} s^{-1}
CH_2=$CHCH_3$	*sec*-$RO_2^{\bullet}$	−2.3	56.2	4.6×10^{-2}	1.4
CH_2=$CHCH_3$	*tert*-$RO_2^{\bullet}$	4.6	59.3	1.5×10^{-2}	0.5
cis-CH_3CH=$CHCH_3$	$HO_2^{\bullet}$	−18.0	49.5	1.0	21
cis-CH_3CH=$CHCH_3$	*sec*-$RO_2^{\bullet}$	−14.5	51.0	0.6	13
cis-CH_3CH=$CHCH_3$	*tert*-$RO_2^{\bullet}$	−7.6	53.9	0.2	27
trans-CH_3CH=$CHCH_3$	$HO_2^{\bullet}$	−16.4	50.2	0.8	17
trans-CH_3CH=$CHCH_3$	*sec*-$RO_2^{\bullet}$	−12.9	51.6	0.5	11
trans-CH_3CH=$CHCH_3$	*tert*-$RO_2^{\bullet}$	−6.0	54.6	0.2	2.0
CH_2=$CHCH_2CH_3$	$HO_2^{\bullet}$	−28.9	45.1	1.7	26
CH_2=$CHCH_2CH_3$	*sec*-$RO_2^{\bullet}$	−24.7	46.8	0.9	15
CH_2=$CHCH_2CH_3$	*tert*-$RO_2^{\bullet}$	−17.8	49.6	0.3	6.7
CH_2=$CHCH(CH_3)_2$	$HO_2^{\bullet}$	−33.6	43.2	1.7	23
CH_2=$CHCH(CH_3)_2$	*sec*-$RO_2^{\bullet}$	−30.1	44.6	1.0	15
CH_2=$CHCH(CH_3)_2$	*tert*-$RO_2^{\bullet}$	−23.2	47.4	0.4	6.5
CH_3CH=$C(CH_3)_2$	$HO_2^{\bullet}$	−20.8	48.3	1.6	30
CH_3CH=$C(CH_3)_2$	*sec*-$RO_2^{\bullet}$	−17.3	49.8	0.9	19
CH_3CH=$C(CH_3)_2$	*tert*-$RO_2^{\bullet}$	−10.4	52.7	0.3	7.9
$(CH_3)_2C$=$C(CH_3)_2$	$HO_2^{\bullet}$	−20.3	48.6	2.8	54
$(CH_3)_2C$=$C(CH_3)_2$	*sec*-$RO_2^{\bullet}$	−16.8	50.1	1.6	34
$(CH_3)_2C$=$C(CH_3)_2$	*tert*-$RO_2^{\bullet}$	−9.9	52.9	0.6	15
cyclo-C_5H_8	$HO_2^{\bullet}$	−30.8	44.3	45	6.6×10^2
cyclo-C_5H_8	*sec*-$RO_2^{\bullet}$	−27.3	45.7	27	4.3×10^2
cyclo-C_5H_8	*tert*-$RO_2^{\bullet}$	−20.4	48.5	10	1.9×10^2
cyclo-C_6H_{10}	$HO_2^{\bullet}$	−31.7	44.0	50	7.2×10^2
cyclo-C_5H_8	*sec*-$RO_2^{\bullet}$	−28.2	45.4	30	4.7×10^2
cyclo-C_5H_8	*tert*-$RO_2^{\bullet}$	−21.3	48.1	11	2.1×10^2
Cyclopentadiene	$HO_2^{\bullet}$	−71.8	29.4	5.0×10^3	2.9×10^4
Cyclopentadiene	*sec*-$RO_2^{\bullet}$	−68.3	30.6	3.2×10^3	2.0×10^4
Cyclopentadiene	*tert*-$RO_2^{\bullet}$	−61.4	33.0	1.3×10^3	9.8×10^3

Hydrocarbon	$RO_2^•$	ΔH/ kJ mol^{-1}	E/ kJ mol^{-1}	k (333 K)/ l mol^{-1} s^{-1}	k (400 K)/ l mol^{-1} s^{-1}
1.3-Cyclohexadiene	$HO_2^•$	−63.6	32.2	3.5×10^3	2.5×10^4
1.3-Cyclohexadiene	*sec*-$RO_2^•$	−60.1	33.4	2.3×10^3	1.7×10^4
1.3-Cyclohexadiene	*tert*-$RO_2^•$	−53.2	35.9	9.3×10^2	8.2×10^3
$C_6H_5CH_3$	$HO_2^•$	−0.8	48.1	0.86	16
$C_6H_5CH_3$	*sec*-$RO_2^•$	2.7	49.7	0.48	9.7
$C_6H_5CH_3$	*tert*-$RO_2^•$	9.6	52.8	0.16	3.8
2-$CH_3C_6H_5CH_3$	$HO_2^•$	1.9	49.3	1.11	22
2-$CH_3C_6H_5CH_3$	*sec*-$RO_2^•$	5.4	50.7	0.67	14
2-$CH_3C_6H_5CH_3$	*tert*-$RO_2^•$	12.3	53.9	0.21	5.5
4-$CH_3C_6H_4CH_3$	$HO_2^•$	−3.6	46.8	2.74	46
4-$CH_3C_6H_4CH_3$	*sec*-$RO_2^•$	−0.1	48.3	1.59	30
4-$CH_3C_6H_4CH_3$	*tert*-$RO_2^•$	6.8	51.4	0.52	12
4-$CH_3OC_6H_4CH_3$	$HO_2^•$	−3.8	46.7	1.42	24
4-$CH_3OC_6H_4CH_3$	*sec*-$RO_2^•$	−0.3	48.2	0.83	15
4-$CH_3OC_6H_4CH_3$	*tert*-$RO_2^•$	6.6	51.3	0.27	6.0
4-$(CH_3)_3CC_6H_4CH_3$	$HO_2^•$	−1.9	47.5	1.06	19
4-$(CH_3)_3CC_6H_4CH_3$	*sec*-$RO_2^•$	1.6	49.0	0.62	12
4-$(CH_3)_3CC_6H_4CH_3$	*tert*-$RO_2^•$	8.5	52.2	0.19	4.6
4-$ClC_6H_4CH_3$	$HO_2^•$	−1.7	47.6	1.02	18
4-$ClC_6H_4CH_3$	*sec*-$RO_2^•$	1.8	49.1	0.60	12
4-$ClC_6H_4CH_3$	*tert*-$RO_2^•$	8.7	52.2	0.19	4.6
4-$CNC_6H_4CH_3$	$HO_2^•$	−7.1	45.2	2.44	38
4-$CNC_6H_4CH_3$	*sec*-$RO_2^•$	−3.6	46.8	1.37	23
4-$CNC_6H_4CH_3$	*tert*-$RO_2^•$	3.3	49.8	0.46	9.4
4-$NO_2C_6H_4CH_3$	$HO_2^•$	0.5	48.6	0.71	14
4-$NO_2C_6H_4CH_3$	*sec*-$RO_2^•$	4.0	50.1	0.42	8.6
4-$NO_2C_6H_4CH_3$	*tert*-$RO_2^•$	10.9	53.3	0.13	3.3
$C_6H_5CH_2CH_3$	$HO_2^•$	−12.0	43.1	3.47	47
$C_6H_5CH_2CH_3$	*sec*-$RO_2^•$	−8.5	44.6	2.02	30

Hydrocarbon	$RO_2^•$	ΔH/ kJ mol^{-1}	E/ kJ mol^{-1}	k (333 K)/ l mol^{-1} s^{-1}	k (400 K)/ l mol^{-1} s^{-1}
$C_6H_5CH_2CH_3$	*tert*-$RO_2^•$	−1.6	47.6	0.68	12
$C_6H_5CH(CH_3)_2$	$HO_2^•$	−18.4	40.4	4.60	53
$C_6H_5CH(CH_3)_2$	*sec*-$RO_2^•$	−14.9	41.9	2.68	34
$C_6H_5CH(CH_3)_2$	*tert*-$RO_2^•$	−8.0	44.8	0.94	14
$(C_6H_5)_2CH_2$	$HO_2^•$	−19.0	40.2	9.89	1.1×10^2
$(C_6H_5)_2CH_2$	*sec*-$RO_2^•$	−15.5	41.7	5.75	72
$(C_6H_5)_2CH_2$	*tert*-$RO_2^•$	−8.6	44.6	2.02	30
$(C_6H_5)_2CHCH_3$	$HO_2^•$	−24.8	37.8	11.81	1.2×10^2
$(C_6H_5)_2CHCH_3$	*sec*-$RO_2^•$	−21.3	39.2	7.10	76
$(C_6H_5)_2CHCH_3$	*tert*-$RO_2^•$	−14.4	42.1	2.50	32
$C_6H_5CH_2CH_2C_6H_5$	$HO_2^•$	−12.8	42.8	7.74	1.0×10^2
$C_6H_5CH_2CH_2C_6H_5$	*sec*-$RO_2^•$	−9.3	44.3	4.50	66
$C_6H_5CH_2CH_2C_6H_5$	*tert*-$RO_2^•$	−2.4	47.3	1.52	27
$(C_6H_5)_3CH$	$HO_2^•$	−26.2	37.3	14.10	1.3×10^2
$(C_6H_5)_3CH$	*sec*-$RO_2^•$	−22.7	38.7	8.51	88
$(C_6H_5)_3CH$	*tert*-$RO_2^•$	−15.8	41.5	3.10	38
$C_6H_5CH_2CH{=}CH_2$	$HO_2^•$	−26.4	37.2	29.2	2.8×10^2
$C_6H_5CH_2CH{=}CH_2$	*sec*-$RO_2^•$	−22.9	38.6	17.6	1.8×10^2
$C_6H_5CH_2CH{=}CH_2$	*tert*-$RO_2^•$	−16.0	41.4	6.41	78.5
Indane	$HO_2^•$	−14.2	42.4	48.0	6.2×10^2
Indane	*sec*-$RO_2^•$	−10.7	43.7	27.9	3.9×10^2
Indane	*tert*-$RO_2^•$	−3.8	46.7	9.5	1.6×10^2
Tetraline	$HO_2^•$	−20.0	39.7	2.4×10^2	2.6×10^3
Tetraline	*sec*-$RO_2^•$	−16.5	41.2	1.4×10^2	1.7×10^3
Tetraline	*tert*-$RO_2^•$	−9.6	44.1	48.4	7.0×10^2

Table 2.2
Rate constants of isomerization and monomolecular decomposition of peroxyl radicals

$RO_2^•$	Products	T / K	k l mol^{-1} s^{-1}	Ref.
$(CH_3)_2C(O_2^•)CH_2CH(CH_3)_2$	$(CH_3)_2C(OOH)CH_2C^•(CH_3)_2$	373	17.7	1
$CH_3CH(O_2^•)CH_2CH_2CH_3$	$CH_3CH(OOH)CH_2C^•HCH_3$	373	0.87	2
$(CH_3)_2C(O_2^•)CH_2CH_2CH(CH3)_2$	$(CH_3)_2C(OOH)CH_2C^•(CH_3)_2$	373	8.0	3
$(CH_3)_2CHCH_2C(O_2^•)(CH_3)CH_2CH(CH_3)_2$	$(CH_2)_2C^•CH_2C(OOH)(CH_3)CH_2CH(CH_3)_2$	373	46	4
$CH_3CH(O_2^•)(CH_2)_{11}CH_3$	$CH_3CH(OOH)CH_2C^•H(CH_2)_9CH_3$	413	1.7×10^2	5
$CH_3CH(O_2^•)(CH_2)_{13}CH_3$	$CH_3CH(OOH)CH_2C^•H(CH_2)_{11}CH_3$	433	1.4×10^2	6
$C_6H_5CH(O_2^•)OCH_2C_6H_5$	$C_6H_5CH(OOH)OC^•HC_6H_5$	303	86	7
$(CH_3)_2C(O_2^•)C(O)CH(CH_3)_2$	$(CH_3)_2C(OOH)C(O))C^•(CH_3)_2$	348	2.7	8
$(C_2H_5C(O)OCH_2)_2(C_2H_5)CCH(O_2^•)OCOC_2H_5$	$C_2H_5C(O)OC^•H(C_2H_5C(O)OCH_2)(C_2H_5)$ $CCH(OOH)OCOC_2H_5$	383	2.3×10^2	9
$(C_2H_5C(O)OCH_2)_3CCH(O_2^•)OCOC_2H_5$	$(C_2H_5C(O)OCH_2)_2(C_2H_5C(O)OC^•H)C$ $CH(OOH)OCO\ C_2H_5$	383	2.0×10^2	9
$CH_3CH_2C(O)OCH_2(CH_3)_2CCH(O_2^•)OCOC_2H_5$	$CH_3CH_2C(O)OC^•H(CH_3)_2CCH(OOH)O$ $COCH_2CH_3$	383	3.4×10^2	9
$CH_3(CH_2)_4CH(O_2^•)CH_2CH(OOH)(CH_2)_7CH_3$	$CH_3(CH_2)_4CH(OOH)CH_2C^•(OOH)(CH_2)_7$ CH_3	433	3.6×10^3	6
$HOCH_2O_2^•$	$CH_2O + HO_2^•$	295	10	10
$HOCH(O_2^•)CH_3$	$CH_3CHO + HO_2^•$	295	52	10
$HOC(O_2^•)(CH_3)_2$	$(CH_3)_2CO + HO_2^•$	295	6.7×10^2	10
cyclo-$C_6H_{10}(O_2^•)OH$	*cyclo*-$C_6H_{10}O + HO_2^•$	299	1.7×10^3	11
$CH(O_2^•)(OH)CH_2OH$	$CHOCH_2OH + HO_2^•$	295	1.9×10^2	12
$CH(O_2^•)(OH)(CHOH)_2CH_2OH$	$CH(OOH)(OH)(CHOH)_2CH_2OH$	295	1.9×10^2	12
$CH(O_2^•)(OH)(CHOH)_3CH_2OH$	$CH(OOH)(OH)(CHOH)_3CH_2OH$	295	2.2×10^2	12
$CH(O_2^•)(OH)(CHOH)_4CH_2OH$	$CH(OOH)(OH)(CHOH)_4CH_2OH$	295	2.1×10^2	12

Table 2.3
Rate constants of addition of peroxyl radicals to double bond of olefins

Oxidizing compound	Peroxyl radical	T/ K	k / $l\ mol^{-1}\ s^{-1}$ or $\log k = A - E/\theta$	Ref.
$CH_2{=}CHCH_3$	$(CH_3)_3CO_2^{\bullet}$	393	0.60	13
$CH_2{=}CH(CH_2)_3CH_3$	$(CH_3)_3CO_2^{\bullet}$	393	0.83	13
$(CH_3)_2C{=}C(CH_3)_2$	$\sim CH_2CH(O_2^{\bullet})C_6H_5$	323	1.3×10^2	14
$(CH_3)_2C{=}C(CH_3)_2$	$(CH_3)_3CO_2^{\bullet}$	393	22	13
$CH_2{=}CHC(CH_3)_3$	$(CH_3)_3CO_2^{\bullet}$	393	0.56	13
$CH_2{=}CHC(CH_3)_3$	$C_6H_5C(O_2^{\bullet})(CH_3)_2$	393	0.66	13
$C_6H_5CH{=}CH_2$	$C_6H_5(CH_3)_2CO_2^{\bullet}$	323	21.1	15
$C_6H_5CH{=}CH_2$	$(CH_3)_3CO_2^{\bullet}$	323	13.8	15
$C_6H_5CH{=}CH_2$	$HO_2^{\bullet}$	323	78.8	15
$C_6H_5CH{=}CH_2$	1,2-*cyclo*-$[CH(O_2^{\bullet})(CH_2)_3]$-$C_6H_4$	303	8.1	16
$C_6H_5CH{=}CH_2$	$\sim CH_2CH(O_2^{\bullet})C_6H_5$	286–333	$7.67 - 35.1/\theta$	17
$4\text{-}CH_3C_6H_4CH{=}CH_2$	$\sim CH_2CH(O_2^{\bullet})C_6H_5$	323	69.0	14
$3\text{-}ClC_6H_4CH{=}CH_2$	$\sim CH_2CH(3\text{-}ClC_6H_4)O_2^{\bullet}$	313	1.0×10^2	18
$4\text{-}ClC_6H_4CH{=}CH_2$	$\sim CH_2CH(O_2^{\bullet})C_6H_5$	323	55.0	14
$4\text{-}ClC_6H_4CH{=}CH_2$	$\sim CH_2CH(4\text{-}ClC_6H_4)O_2^{\bullet}$	313	1.2×10^2	18
$4\text{-}FC_6H_4CH{=}CH_2$	$\sim CH_2CH(O_2^{\bullet})C_6H_5$	323	55.7	14
$4\text{-}BrC_6H_4CH{=}CH_2$	$\sim CH_2CH(O_2^{\bullet})C_6H_5$	323	55.7	14
$4\text{-}IC_6H_4CH{=}CH_2$	$\sim CH_2CH(O_2^{\bullet})C_6H_5$	323	70	14
$4\text{-}CNC_6H_4CH{=}CH_2$	$\sim CH_2CH(O_2^{\bullet})C_6H_5$	323	44.5	14
$3\text{-}NO_2C_6H_4CH{=}CH_2$	$\sim CH_2CH(O_2^{\bullet})C_6H_5$	323	39.5	14
$4\text{-}NO_2C_6H_4CH{=}CH_2$	$\sim CH_2CH(O_2^{\bullet})C_6H_5$	323	40.8	14
$4\text{-}CNC_6H_4CH{=}CH_2$	$\sim CH_2CH(4\text{-}CNC_6H_4)O_2^{\bullet}$	313	91.0	18
$C_6H_5C(CH_3){=}CH_2$	$\sim CH_2C(CH_3)(C(O)OCH_3)O_2^{\bullet}$	323	25.2	15
$C_6H_5C(CH_3){=}CH_2$	$\sim CH_2C(CH_3)(O_2^{\bullet})C_6H_5$	286–323	$6.82 - 33.7/\theta$	19
$C_6H_5C(CH_3){=}CH_2$	$\sim CH_2CH(O_2^{\bullet})C_6H_5$	323	2.1×10^2	15
$C_6H_5C(CH_3){=}CH_2$	$(CH_3)_3CO_2^{\bullet}$	323	17.2	15

Oxidizing compound	Peroxyl radical	T/ K	k / $l\ mol^{-1}\ s^{-1}$ or $\log k = A - E/\theta$	Ref.
$C_6H_5C(CH_3){=}CH_2$	$C_6H_5(O_2^\bullet)(CH_3)_2$	323	34	15
$CH_3CH{=}CHC_6H_5$	$CH(CH_3)CH(C_6H_5)O_2^\bullet$	303	51	19
$CH_3CH{=}CHC_6H_5$	$(CH_3)_3CO_2^\bullet$	303	2.5	20
$C_6H_4(CH)_2C_6H_4$	$C_6H_5CH(O_2^\bullet)CH_3$	323–358	6.03 – 18.6/ θ	21
trans-$C_6H_5CH{=}CHC_6H_5$	$(CH_3)_3CO_2^\bullet$	303	0.44	20
$C_6H_5CH{=}CHC_6H_5$	$HO_2^\bullet$	323	1.1×10^2	15
$C_6H_4(CH)_2C_6H_4$	$C_6H_5CH_2O_2^\bullet$	323–353	13.09 – 60.6/ θ	21
$CH_2{=}C(C_6H_5)_2$	$(CH_3)_3CO_2^\bullet$	303	9.2	20
1,2-*cyclo*-[$CH{=}CHCH_2$]-C_6H_4	1,2-*cyclo*-[~$CHCH(O_2^\bullet)CH_2$]-C_6H_4	303	1.2×10^2	19
1,2-*cyclo*-[$CH{=}CHCH_2CH_2$]-C_6H_4	1,2-*cyclo*-[~$CHCH(O_2^\bullet)(CH_2)_2$]-$C_6H_4$	303	2.9×10^2	22
cyclo-[$C(C_6H_5){=}CH(CH_2)_4$]	*cyclo*-[$C(O_2^\bullet)(C_6H_5)C^\bullet H(CH_2)_4$	323–343	8.15 – 35.1/ θ	23
cyclo-[$C(C_6H_5){=}CH(CH_2)_4$)]	*cyclo*-[$C(O_2^\bullet)(C_6H_5)C^\bullet H(CH_2)_4$	343	6.6×10^2	23
cyclo-[$C(C_6H_5){=}CH(CH_2)_4$]	*cyclo*-[$C(O_2^\bullet)(C_6H_5)C^\bullet H(CH_2)_4$	333	4.9×10^2	23
cyclo-[$C(C_6H_5){=}CH(CH_2)_4$]	*cyclo*-[$C(O_2^\bullet)(C_6H_5)C^\bullet H(CH_2)_4$	323	2.8×10^2	23
$CH_2{=}CHCN$	~$CH_2CH(O_2^\bullet)CN$	323	10.8	15
$CH_2{=}CHCN$	~$CH_2CH(O_2^\bullet)CN$	303	3.25	20
$CH_2{=}CHCN$	~$CH_2CH(O_2^\bullet)C_6H_5$	323	24.5	15
$CH_2{=}CHCN$	$HO_2^\bullet$	323	35.0	15
$CH_2{=}CHCN$	$(CH_3)_3CO_2^\bullet$	323	0.70	15
$CH_2{=}CHCN$	$C_6H_5(CH_3)_2CO_2^\bullet$	323	1.5	15
$CH_3CH{=}CHCN$	~$CH(CH_3)CH(CN)O_2^\bullet$	303	4.5	20
$CH_2{=}CHOC(O)CH_3$	$C_6H_5(CH_3)_2CO_2^\bullet$	323	0.20	15
$CH_2{=}CHOC(O)CH_3$	$(CH_3)_3CO_2^\bullet$	323	0.10	15
$CH_2{=}CHOC(O)CH_3$	$HO_2^\bullet$	323	6.8	15
$CH_2{=}CHOCH_2CH_3$	$(CH_3)_3CO_2^\bullet$	303	4.0×10^{-2}	20
$CH_3CH_2C(O)CH{=}CH_2$	$(CH_3)_3CO_2^\bullet$	303	1.3×10^{-2}	20
$(CH_3)_2C{=}CHC(O)CH_3$	$HO_2^\bullet$	323	8.7	15
$CH_2{=}CHC(O)CH_2CH_3$	~$CH_2CH(C(O)CH_2CH_3)O_2^\bullet$	303	2.7	20
$CH_2{=}CHC(O)OCH_3$	$C_6H_5(CH_3)_2CO_2^\bullet$	323	0.50	15

Oxidizing compound	Peroxyl radical	T/ K	k / l mol^{-1} s^{-1} or log k = A − E/ θ	Ref.
CH_2=CHC(O)OCH_3	$(CH_3)_3CO_2^{\bullet}$	323	0.40	15
CH_2=CHC(O)OCH_3	~CH_2CH($O_2^{\bullet}$)C_6H_5	323	9.8	15
CH_2=CHC(O)OCH_3	~CH_2CH($O_2^{\bullet}$)C(O)OCH_3	323	1.7	15
CH_2=CHC(O)OCH_3	~CH_2CH($O_2^{\bullet}$)OC(O)CH_3	323	3.4	15
CH_2=CHOC(O)CH_3)	~CH_2CH($O_2^{\bullet}$)OC(O)CH_3	323	2.8	15
CH_2=C(CH_3)C(O)OCH_3	$C_6H_5(CH_3)_2CO_2^{\bullet}$	323	1.8	15
CH_2=C(CH_3)C(O)OCH_3	$(CH_3)_3CO_2^{\bullet}$	323	1.1	15
CH_2=C(CH_3)C(O)OCH_3	$HO_2^{\bullet}$	323	40	15
CH_2=C(CH_3)C(O)OCH_3	~CH_2CH($O_2^{\bullet}$)C_6H_5	323	12	15
CH_2=C(CH_3)C(O)OCH_3	~CH_2C(CH_3)($O_2^{\bullet}$)C(O)CH_3	303–323	8.92 − 53.5/ θ	24
$(CH_3)_2$C=CHOC(O)CH_3	$(CH_3)_3CO_2^{\bullet}$	303	1.0 × 10^{-2}	20
$(CH_3)_2$C=CHC(O)OCH_3	~C$(CH_3)_2$CH(C(O)OCH_3)$O_2^{\bullet}$	303	0.20	20
trans-CH_3CH=CHC(O)OCH_3	$HO_2^{\bullet}$	323	13.1	15
trans-CH_3CH=CHC(O)OCH_2CH_3	$HO_2^{\bullet}$	323	14.2	15
cis-CH_3CH_2OC(O)CH=CHC(O)OCH_2 CH_3	$HO_2^{\bullet}$	323	1.4	15
trans-CH_3CH_2OC(O)CH=CHC(O)OCH_2 CH_3	$HO_2^{\bullet}$	323	0.60	15
CH_2=C(CH_3)C(O)O$(CH_2)_3CH_3$	~CH_2C($O_2^{\bullet}$)(CH_3)C(O)$(CH_2)_3CH_3$	303–323	7.80 − 45.6/ θ	24
CH_2=CHC(O)O$(CH_2)_3CH_3$	~CH_2CH(C(O)O$(CH_2)_3CH_3$)$O_2^{\bullet}$	323	1.4	24
CH_2=C(CH_3)C(O)OCH_2CH$(CH_3)_2$	~CH_2C(CH_3)(C(O)OCH_2CH$(CH_3)_2$$O_2^{\bullet}$	323	2.4	24
[(CH_2=CHC(O)OCH_2]$_2$C$(CH_3)_2$	$C_6H_5(CH_3)_2CO_2^{\bullet}$	323	0.76	25
[CH_2=CHC(O)OCH_2]$_4$C	$C_6H_5(CH_3)_2CO_2^{\bullet}$	323	1.28	25
[CH_2=C(CH_3)C(O)OCH_2]$_4$C	$C_6H_5(CH_3)_2CO_2^{\bullet}$	323	3.08	25
CH_2=CHC(O)NH_2	$C_6H_5(CH_3)_2CO_2^{\bullet}$	323	0.40	15
CH_2=CHC(O)NH_2	$(CH_3)_3CO_2^{\bullet}$	323	0.20	15
CH_2=C(CH_3)C(O)NH_2	$(CH_3)_3CO_2^{\bullet}$	323	0.30	15

Table 2.4
Rate constants of addition of alkyl radicals to molecular oxygen

$R^•$	T/ K	Solvent	k / $l\ mol^{-1}\ s^{-1}$	Ref.
$C^•H_3$	296	H_2O	4.7×10^9	26
$CH_3C^•H_2$	298	RH	2.9×10^9	27
$CH_3C^•H(CH_2)_7CH_3$	298	RH	4.8×10^9	28
$CH_3(CH_2)_{13}C^•HCH_3$	298	RH	1.5×10^9	29
$CH_3(CH_2)_{14}C^•HCH_3$	298	RH	1.5×10^9	29
$CH_3(CH_2)_4CH{=}CHC^•HCH{=}CH(CH_2)_7COOH$	295	RH	3.0×10^8	30
$CH_3CH_2(CH{=}CHCH_2)_2CH{=}CHC^•H(CH_2)_6C(O)OH$	295	RH	3.0×10^8	30
cyclo-$[CH{=}CHC^•HCH{=}CHCH_2]$	300	RH	1.6×10^9	31
cyclo-$[CH(OH)C[C(CH_3)_3]{=}CHC^•(CH_3)CH{=}C(C(CH_3)_3]$	298	RH	9.0×10^7	29
$HOCH{=}CH^•$	295	RH	1.0×10^9	32
$C_6H_5C^•H_2$	294	RH	2.0×10^9	33
$C_6H_5C^•HCH_3$	323	RH	8.8×10^8	34
$C_6H_5(CH_3)_2C^•$	323	RH	9.0×10^8	31
$(C_6H_5)_2C^•H$	294	RH	7.5×10^8	33
$(C_6H_5)_3C^•$	293	RH	1.2×10^9	35
$RCH_2C^•HC_6H_5$	323	RH	8.0×10^7	34
$4\text{-}NO_2C_6H_4C^•H_2$	294	RH	9.0×10^8	33
$(C_6H_5)_2NC^•HCH_3$	300	RH	4.9×10^9	31

Table 2.5
Rate constants of recombination and disproportionation of peroxyl radical in hydrocarbon solutions

Peroxyl radical	T/ K	k/ $l\ mol^{-1}\ s^{-1}$ or $\log k = A - E/\theta$	Ref.
$(CH_3)_2CHO_2^•$	210–300	$7.89 - 10.34/\theta$	36
$CH_3(CH_2)_2CH_2O_2^•$	303	4×10^7	37
$CH_3CH_2CH(O_2^•)CH_3$	193–257	$9.0 - 11.3/\theta$	38

Peroxyl radical	T/ K	k/ l mol^{-1} s^{-1} or $\log k = A - E/\theta$	Ref.
$(CH_3)_3CO_2^{\bullet}$	193–257	9.3 – 31.8/θ	38
$CH_3CH(O_2^{\bullet})(CH_2)_2CH_3$	233–310	8.5 – 10.9/θ	67
$CH_3(CH_2)_2C(O_2^{\bullet})(CH_3)_2$	213–273	11.1 – 39/θ	40
$CH_3CH(O_2^{\bullet})(CH_2)_3CH_3$	283–320	7.46 – 8.4/θ	41
$CH_3CH(O_2^{\bullet})(CH_2)_4CH_3$	294–324	7.46 – 8.4/θ	41
$(CH_3)_3CC(O_2^{\bullet})(CH_3)_2$	243–293	9.2 – 31.4/θ	40
$CH_3CH(O_2^{\bullet})(CH_2)_5CH_3$	283–356	7.46 – 8.4/θ	41
$CH_3CH(O_2^{\bullet})(CH_2)_6CH_3$	283–324	7.46 – 8.4/θ	41
$CH_3CH(O_2^{\bullet})(CH_2)_7CH_3$	283–355	7.46 – 8.4/θ	41
$CH_3CH(O_2^{\bullet})(CH_2)_9CH_3$	284–355	7.46 – 8.4/θ	41
$(CH_3)_2C(O_2^{\bullet})CH_2CH(CH_3)CH_2CH(CH_3)CH(CH_3)_2$	243–293	13.41 – 46/θ	42
$CH_3CH(O_2^{\bullet})(CH_2)_{10}CH_3$	293–358	7.46 – 8.4/θ	41
cyclo-$[CH(O_2^{\bullet})(CH_2)_4]$	175–200	9.6 – 10.9/θ	43
cyclo-$[CH(O_2^{\bullet})(CH_2)_5]$	285–333	7.29 – 5.4/θ	41
cyclo-$[CH(O_2^{\bullet})(CH_2)_6]$	298	8.6×10^6	44
cyclo-$[CH(O_2^{\bullet})(CH_2)_7]$	298	1.4×10^7	44
cyclo-$[CH(O_2^{\bullet})(CH_2)_{11}]$	345–417	8.12 – 7.8/θ	46
$CH_2{=}CHCH(O_2^{\bullet})CH{=}CH_2$	303	1.1×10^9	68
$(CH_3)_2C{=}CHCH(O_2^{\bullet})CH_3$	313–333	7.4 – 5.0/θ	47
$CH_3CH_2CH{=}CHCH(O_2^{\bullet})CH_2CH_3$	303	6.4×10^6	37
cis-$CH_3(CH_2)_2CH{=}CHCH(O_2^{\bullet})CH_2CH_3$	323	1.5×10^7	48
$C_6H_5CH{=}CHCH_2O_2^{\bullet}$	303	4.4×10^8	49
$CH_2{=}CHCH(O_2^{\bullet})(CH_2)_4CH_3$	298	1.0×10^7	50
$(CH_3)_2C{=}CHC(O_2^{\bullet})(CH_3)CH_2CH(CH_3)CH(CH_3)_2$	234–293	10.2 – 21/θ	42
$CH_3CH_2CH{=}CHCH(O_2^{\bullet})CH{=}CHCH_2CH{=}CHCH_2C(O)OCH_3$	353	1.8×10^7	51
$CH_3(CH_2)_4(CH{=}CH)_2CH(O_2^{\bullet})(CH_2)_7C(O)OCH_3$	303	2.43×10^6	52
$CH_3(CH_2)_7CH{=}CHCH(O_2^{\bullet})(CH_2)_6C(O)OCH_3$	303	1.06×10^6	22
$CH_3(CH_2)_4CH{=}CHCH(O_2^{\bullet})CH{=}CH(CH_2)_7C(O)OCH_2CH_3$	298	3.0×10^7	53
cyclo-$[CH(O_2^{\bullet})CH{=}CH(CH_2)_2]$	193–257	7.8 – 4.2/θ	38

Peroxyl radical	*T*/ K	*k*/ l mol^{-1} s^{-1} or log *k* = *A* – *E*/θ	Ref.
$C_6H_5CH_2O_2^•$	303	3.0 × 10^8	22
cyclo-[CH=CHCH($O_2^•$)(CH_2)$_3$]	282–319	8.75 – 8.3/θ	54
4-CH_3-$C_6H_4CH_2O_2^•$	348	2.6 × 10^8	55
cyclo-[CH($O_2^•$)C(CH_3)=CH(CH_2)$_3$]	313	8.6 × 10^5	56
cyclo-[CH=CHCH($O_2^•$)CH=CHCH$_2$]	273	1.04 × 10^8	57
C_6H_5CH($O_2^•$)CH_3	323–353	9.1 – 9.4/θ	58
4-CH_3-C_6H_4CH($O_2^•$)CH_3	348	1.02 × 10^7	59
cyclo-[CH($O_2^•$)C(C_6H_5)=CH(CH_2)$_3$]	313–343	2.5 × 10^8	23
3-CH_2CH_3-C_6H_4CH($O_2^•$)CH_3	348	1.1 × 10^7	60
4-CH_2CH_3-C_6H_4CH($O_2^•$)CH_3	348	1.0 × 10^7	61
3.5-(CH_2CH_3)$_2$-C_6H_3CH($O_2^•$)CH_3	348	1.0 × 10^7	61
2,4,5-(CH_2CH_3)$_3$-C_6H_2CH($O_2^•$)CH_3	348	1.9 × 10^6	61
3,4,5-(CH_2CH_3)$_3$-C_6H_2CH($O_2^•$)CH_3	348	3.3 × 10^6	61
2,3,4,5-(CH_2CH_3)$_4$-C_6HCH($O_2^•$)CH_3	348	6.0 × 10^5	61
2,3,4,5,6-(CH_2CH_3)$_5$-C_6CH($O_2^•$)CH_3	348	4.5 × 10^5	61
3-OCH_3-C_6H_4CH($O_2^•$)CH_3	348	1.3 × 10^7	60
4-OCH_3-C_6H_4CH($O_2^•$)CH_3	348	9.0 × 10^6	60
4-Cl-C_6H_4CH($O_2^•$)CH_3	348	1.6 × 10^7	60
4-C(O)CH_3-C_6H_4CH($O_2^•$)CH_3	348	1.5 × 10^7	60
4-NO_2-C_6H_4 CH($O_2^•$)CH_3	348	2.3 × 10^7	60
4-NO_2-C_6H_4 CH($O_2^•$)CH_3	348	2.4 × 10^7	60
C_6H_5CH($O_2^•$)CH_2CH_3	348	3.0 × 10^7	59
C_6H_5C($O_2^•$)(CH_3)$_2$	164–243	9.2 – 25.1/θ	62
C_6H_5C($O_2^•$)(CH_3)CH_2CH_3	303–329	9.30 – 37.66/θ	64
4-CH_3-C_6H_4C($O_2^•$)(CH_3)$_2$	308–338	2.0 × 10^5	65
3-OCH_3-C_6H_4C($O_2^•$)(CH_3)$_2$	303	6.0 × 10^4	68
4-OCH_3-C_6H_4C($O_2^•$)(CH_3)$_2$	303	4.0 × 10^4	63
4-C(O)OCH_3-C_6H_4C($O_2^•$)(CH_3)$_2$	303	3.0 × 10^4	63
4-CH_3-C_6H_4C($O_2^•$)(CH_3)CH_2CH_3	348	6.6 × 10^6	59

Peroxyl radical	T/ K	k/ l mol^{-1} s^{-1} or $\log k = A - E/\theta$	Ref.
4-$CH(CH_3)_2$-$C_6H_4C(O_2^{\bullet})(CH_3)_2$	348	1.1×10^5	66
3-$CH(CH_3)_2$-$C_6H_4C(O_2^{\bullet})(CH_3)_2$	348	1.0×10^5	66
3-$CH(CH_3)_2$-$C_6H_4C(O_2^{\bullet})(CH_3)_2$	348	1.0×10^5	67
3,5-$[CH(CH_3)_2]_2$-$C_6H_3C(O_2^{\bullet})(CH_3)_2$	348	1.0×10^5	66
2,4-$[CH(CH_3)_2]_2$-$C_6H_3C(O_2^{\bullet})(CH_3)_2$	348	3.2×10^5	66
2,4,5-$[CH(CH_3)_2]_3$-$C_6H_2C(O_2^{\bullet})(CH_3)_2$	348	8.0×10^6	67
$(C_6H_5)_2CHO_2^{\bullet}$	303	1.6×10^8	49
$C_6H_5C(O_2^{\bullet})(CH_2CH_3)_2$	348	2.7×10^6	59
cyclo-[1,2-*cyclo*-$[CH(O_2^{\bullet})(CH_2)_3]$-$C_6H_4$	286–323	$9.94 - 18.0/\theta$	49
cyclo-$[(CH_2)_2C(O_2^{\bullet})(C_6H_5)]$	348	1.9×10^6	59
cyclo-$[(CH_2)_3C(O_2^{\bullet})(C_6H_5)]$	348	8×10^3	59
cyclo-$[(CH_2)_4C(O_2^{\bullet})(C_6H_5)]$	348	8.1×10^5	59
cyclo-$[(CH_2)_5C(O_2^{\bullet})(C_6H_5)]$	348	8.9×10^5	59

Table 2.6
Rate constants of disproportionation of two peroxyl radicals of different structure

$R_1O_2^{\bullet}$	$R_2O_2^{\bullet}$	T/ K	Solvent	k/ l mol^{-1} s^{-1}	Ref.
$CH_3O_2^{\bullet}$	$(CH_3)_3CO_2^{\bullet}$	293		1.8×10^6	68
$CH_3(CH_2)_2CH(O_2^{\bullet})O(CH_2)_3CH_3$	$C_6H_5C(CH_3)_2O_2^{\bullet}$	333	R_1H / R_2H	2.9×10^6	69
$C_6H_5C(CH_3)_2O_2^{\bullet}$	$C_6H_5CH_2O_2^{\bullet}$	348	R_1H / R_2H	5.7×10^6	70
$(CH_3)_2CH(CH_2)_4C(CH_3)_2O_2^{\bullet}$	$C_6H_5CH(CH_3)O_2^{\bullet}$	350	R_1H / R_2H	1.2×10^6	71
$C_6H_5C(CH_3)_2O_2^{\bullet}$	$(C_6H_5)_2CHO_2^{\bullet}$	348	R_1H / R_2H	2.8×10^6	70
$C_6H_5C(CH_3)_2O_2^{\bullet}$	$C_6H_5CH(CH_3)O_2^{\bullet}$	348	R_1H / R_2H	1.5×10^6	70
$C_6H_5C(CH_3)_2O_2^{\bullet}$	4-$CH_3C_6H_4CH_2O_2^{\bullet}$	348	R_1H / R_2H	2.7×10^6	70
$C_6H_5C(CH_3)_2O_2^{\bullet}$	$C_6H_5C(CH_3)(CH_2CH_3)O_2^{\bullet}$	348	R_1H / R_2H	8.8×10^4	70
$C_6H_5C(CH_3)_2O_2^{\bullet}$	$(C_6H_5)_2C(CH_3)O_2^{\bullet}$	348	R_1H / R_2H	5.8×10^6	70
$(C_6H_5)_2C(CH_3)O_2^{\bullet}$	$(C_6H_5)_2CHO_2^{\bullet}$	348	R_1H / R_2H	4.1×10^7	70

$R_1O_2^•$	$R_2O_2^•$	T/ K	Solvent	k/ $l\ mol^{-1}\ s^{-1}$	Ref.
$(C_6H_5)_2C(CH_3)O_2^•$	$C_6H_5CH_2O_2^•$	348	R_1H / R_2H	1.3×10^8	70
$C_6H_5C(CH_3)_2O_2^•$	1,2-*cyclo*-$[CH(O_2^•)(CH_2)_3]$-C_6H_4	353	R_1H / R_2H	6.2×10^6	72
$C_6H_5C(CH_3)_2O_2^•$	4-$CH(CH_3)_2C_6H_4C(CH_3)_2O_2^•$	348	R_1H / R_2H	9.0×10^4	67
$C_6H_5C(OH)(CH_3)O_2^•$	1,2-*cyclo*-$[CH(O_2^•)(CH_2)_3]$-C_6H_4	333	R_1H / R_2H	1.4×10^7	73
$C_6H_5C(CH_3)_2O_2^•$	$C_6H_5CH(CH_2CH_3)O_2^•$	348	R_1H / R_2H	1.8×10^6	70
$C_6H_5C(CH_3)_2O_2^•$	$C_6H_5CH(O_2^•)C(CH_3)_3$	348	R_1H / R_2H	3.9×10^6	70
$C_6H_5C(CH_3)_2O_2^•$	*cyclo*-$[C(O_2^•)(C_6H_5)(CH_2)_5]$	348	R_1H / R_2H	8.1×10^5	70
$C_6H_5C(CH_3)_2O_2^•$	$C_6H_5CH(O_2^•)(CH_2)_4CH_3$	348	R_1H / R_2H	6.9×10^6	70
$C_6H_5CH(CH_3)O_2^•$	$C_6H_5C(O_2^•)(CH_3)CH_2CH_3$	348	R_1H / R_2H	1.9×10^6	74
$C_6H_5CH(CH_3)O_2^•$	$C_6H_5CH_2O_2^•$	348	R_1H / R_2H	2.9×10^8	70
$(C_6H_5)_2CHO_2^•$	$C_6H_5CH(OH)O_2^•$	348	R_1H / R_2H	2.9×10^7	70
$C_6H_5CH(CH_3)O_2^•$	$CH_2CH(C_6H_5)O_2^•$	338	R_1H / R_2H	3.6×10^8	75
$(C_6H_5)_2CHO_2^•$	$C_6H_5CH_2O_2^•$	348	R_1H / R_2H	3.1×10^8	70
$C_6H_5CH(CH_3)O_2^•$	$C_6H_5CH(OH)O_2^•$	348	R_1H / R_2H	2.8×10^7	70
$C_6H_5C(CH_3)_2O_2^•$	$(CH_3)_2C(O_2^•)OCH(CH_3)_2$	333	R_1H / R_2H	1.8×10^5	69
$(CH_3)_2C(O_2^•)OCH(CH_3)_2$	*cyclo*-$[CH{=}CH(O_2^•)(CH_2)_3]$	333	R_1H / R_2H	2.0×10^5	69
cyclo-$[OCH(O_2^•)(CH_2)_3]$	*cyclo*-$[CH{=}CHCH(O_2^•)(CH_2)_3]$	333	R_1H / R_2H	1.0×10^7	69
cyclo-$[OCH(O_2^•)(CH_2)_3]$	1,2-*cyclo*-$[CH(O_2^•)(CH_3)_3]$-C_6H_4	333	R_1H / R_2H	1.2×10^7	69
cyclo-$[CH{=}CHCH(O_2^•)(CH_2)_3]$	*cyclo*-$[OCH(O_2^•)(CH_2)_3]$	333	R_1H / R_2H	1.1×10^7	76
cyclo-$[CH{=}CHCH(O_2^•)(CH_2)_3]$	$CH_3(CH_2)_2CH(O_2^•)O(CH_2)_3CH_3$	333	R_1H / R_2H	8.2×10^6	76
$C_6H_5C(OH)(CH_3)O_2^•$	*cyclo*-$[C(OH)(O_2^•)(CH_2)_5]$	333	R_1H / R_2H	2.5×10^6	73
~$CH_2CH(C_6H_5)O_2^•$	~$CH_2CH(CN)O_2^•$	323	R_1H / R_2H	4.8×10^7	77
~$CH_2CH(C_6H_5)O_2^•$	~$CH_2C(CH_3)(O_2^•)C(O)OCH_3$	323	R_1H / R_2H	2.5×10^6	77
~$CH_2CH(C_6H_5)O_2^•$	~$CH_2CH(O_2^•)C(O)OCH_3$	323	R_1H / R_2H	1.5×10^6	77
~$CH_2C(CH_3)(C_6H_5)O_2^•$	~$CH_2C(CH_3)(O_2^•)(O)OCH_3$	323	R_1H / R_2H	2.7×10^5	77
~$CH_2C(CH_3)(C_6H_5)O_2^•$	~$CH_2CH(O_2^•)(O)OCH_3$	323	R_1H / R_2H	6.9×10^4	77

Table 2.7
Rate constants of decomposition of azo-compounds into free radicals in liquid phase

RN=NR	T/ K	log (A/ s^{-1})	E/ kJ mol^{-1}	Ref.
$[(CH_3)_3C]_2N_2$	373–473	17.16	183.7	78
$[CH_3CH_2CH_2(CH_3)_2C]_2N_2$	433–473	16.38	173.2	78
$[(CH_3)_3C(CH_3)_2C]_2N_2$	443–473	15.06	159.3	79
$[(CH_3)_3CCH_2(CH_3)_2C]_2N_2$	403–433	15.57	133.5	79
$[C_6H_5CH_2]_2N_2$	418–448	14.45	145.8	79
[p-CH_3-$C_6H_4CH_2]_2N_2$	418–448	14.77	148.2	79
[p-CH_3O-$C_6H_4CH_2]_2N_2$	418–448	14.72	147.0	79
[p-Cl-$C_6H_4CH_2]_2N_2$	418–448	15.49	153.0	79
[p-C_6H_5-$C_6H_4CH_2]_2N_2$	408–438	14.05	140.3	79
$C_6H_5CH(CH_3)N{=}NCH(CH_3)_2$	416–438	15.41	152.7	80
$C_6H_5C(CH_3)_2N{=}NCH(CH_3)_2$	393–416	16.45	153.6	80
$[C_6H_5CH_2C(CH_3)_2]_2N_2$	423	14.21	152.3	81
$C_6H_5N{=}NCH(C_6H_5)_2$	397–417	14.23	142.3	82
$C_6H_5N{=}NC(C_6H_5)_3$	318–328	16.04	122.6	83
$[(C_6H_5)_2CH]_2N_2$	327–337	13.78	111.3	84
$[C_6H_5C(CH_3)_2]N_2$	314–331	15.41	122.6	85
cyclo-$[N_2CH(C_6H_5)CH_2CH(C_6H_5)]$	333–363	13.84	114.4	79
cis-cyclo-$[N_2CH(C_6H_5)CH_2CH_2CH_2CH(C_6H_5)]$	334–373	15.04	124.3	86
cyclo-$[N_2CH(C_6H_5)CH_2CH_2CH_2CH_2CH(C_6H_5)]$	416–446	15.25	153.6	87
cyclo-$[(C_3H_5)_3C)]_2N_2$	378–408	16.18	146.9	88
$[(CH_3)_2CH$-*cyclo*-$C_3H_5]_2N_2$	398–420	17.20	162.3	88
$[(CH_3CH_2)_2(CN)C]_2N_2$	353–373	17.09	143.1	89
$[(CH_3CH_2CH_2)_2(CN)C]_2N_2$	353–373	15.48	128.9	89
$[(CH_3)_2(CN)C]_2N_2$	343–373	16.84	139.7	90
$[CH_3CH_2(CH_3)(CN)C]_2N_2$	343–353	15.14	129.7	91
$[CH_3CH_2CH_2(CH_3)(CN)C]_2N_2$	343–353	16.65	138.1	91
$[(CH_3)_2CH(CH_3)(CN)C]_2N_2$	343–353	15.83	133.9	91

RN=NR	*T*/ K	log (*A*/ s^{-1})	*E* / kJ mol^{-1}	Ref.
$[(CH_3)_2CHCH_2(CH_3)(CN)C]_2$	343–353	14.74	121.3	91
cyclo-$[(CH_2)_3C(CN)]_2N_2$	393–415	13.12	134.3	92
cyclo-$[(CH_2)_4C(CN)]_2N_2$	353–368	16.78	141.4	92
cyclo-$[(CH_2)_6C(CN)]_2N_2$	322–340	14.10	115.1	92
cyclo-$[(CH_2)_7C(CN)]_2N_2$	310–323	13.97	108.4	92
cyclo-$[(C_3H_5)_2(CN)C]_2N_2$	288–318	13.59	101.7	93
$[C_6H_5C(CH_3)(CN)]_2N_2$	288–306	12.28	84.5	94
$[CH_3O(CH_3)_2C]_2N_2$	423–457	17.62	178.7	79
$[(CH_3)_2(COOCH_3)C]_2N_2$	343–373	14.11	122.6	90
$[CH_3C(O)(CH_3)_2C]_2N_2$	463–490	15.27	171.1	79
$[C_6H_5CH_2O]_2N_2$	303–323	12.23	95.8	94
$[C_6H_5OC(CH_3)_2]_2N_2$	423	12.76	138.9	81
$[HOOCCH_2CH_2(CH_3)(CN)C]_2N_2$	288–323	12.16	108.8	95
$[HOC(O)CH_2CH_2(CN)(CH_3)C]_2N_2$	338–353	15.77	132.6	96
$[HO(O)CCH(CH_3)CH_2(CN)(CH_3)C]_2N_2$	338–353	15.59	121.7	96
$[C_6H_5C(O)C(CH_3)_2]_2N_2$	343–363	15.69	128.1	97

Table 2.8
Rate constants of decomposition of peroxides into free radicals in liquid phase

Peroxide	Solvent	*T*/ K	log (*A*/ s^{-1})	*E*/ kJ mol^{-1}	Ref.
$(CH_3CH_2O)_2$	C_6H_6 / $C_6H_5CH=CH_2$	333–353	14.18	147.3	98
$[(CH_3)_2CHO]_2$	C_6H_6 / $C_6H_5CH=CH_2$	333–353	15.26	156.1	98
$[CH_3(CH_2)_3O]_2$	C_6H_6 / $C_6H_5CH=CH_2$	333–353	13.98	143.5	99
$[(CH_3)(C_2H_5)CHO]_2$	C_6H_6 / $C_6H_5CH=CH_2$	333–353	13.56	142.3	98
$[(CH_3)_3CO]_2$	C_6H_6	333–398	14.50	146.4	100
$[(CH_3)_2(C_2H_5)CO]_2$	$C_6H_5CH=CH_2$	353–383	16.19	157.7	101
$[CH_3(C_2H_5)_2CO]_2$	$C_6H_5CH=CH_2$	353–383	16.55	159.0	101

Peroxide	Solvent	T/ K	log (A/ s^{-1})	E/ kJ mol^{-1}	Ref.
$[C_6H_5(CH_3)_2CO]_2$	C_6H_6	368–423	14.45	142.9	100
$[(CH_3)_3SiO]_2$	C_7H_{16}	456–476	15.04	172.4	102
$(CH_3)_3COOC(CH_3)_2C_2H_5$	$C_{10}H_{22}$	383–403	12.08	126.8	103
$(CH_3)_3COOC(CH_3)_2C_6H_5$	$C_{10}H_{22}$	383–403	15.40	151.9	103
$(CH_3)_3COOCH_2OCH_2OCH_3$	$C_6H_5CH{=}CH_2$	353–383	12.77	130.1	101
$(CH_3)_3COOSi(CH_3)_3$	C_7H_{16}	456–476	15.04	172.4	102
$[CH_3C(O)O]_2$	$(CH_3)_3CCH_2CH(CH_3)_2$	328–358	15.82	134.7	104
$CH_3C(O)OOC(CH_3)_3$	$C_{11}H_{24}$	363–393	14.10	133.5	105
$[CH_3CH_2OC(O)O]_2$	$(CH_2{=}CHCH_2OC(O)CH_2CH_2)_2O$	313–333	15.89	126.2	106
$[(CH_3CH_2C(O)O]_2$	$C_6H_5CH_2CH_3$	347–358	14.75	126.4	107
$[(CH_3)_2CHC(O)O]_2$	CCl_4	318–333	11.74	94.4	108
$[CH_3CH_2CH_2C(O)O]_2$	$C_6H_5CH_2CH_3$	347–358	14.64	124.7	107
$[CH_3(CH_2)_3C(O)O]_2$	$C_6H_5CH_2CH_3$	347–358	14.50	123.8	107
cyclo-$[C_6H_{11}OC(O)O]_2$	CCl_4	308–333	13.21	102.5	109
$[C_6H_5C(O)O]_2$	$C_6H_5CH{=}CH_2$	338–358	16.00	137	110
$[C_6H_5(CH_2)_4C(O)O]_2$	$C_6H_5CH{=}CH_2/CCl_4$	343–358	15.25	130.1	111
$[4\text{-}CH_3\text{-}C_6H_4C(O)O]_2$	$C_6H_5OC(O)CH_3$	346–368	14.30	125.1	112
$[4\text{-}CH_3O\text{-}C_6H_4C(O)O]_2$	Dioxane	333–353	14.02	121.3	113
$[4\text{-}Cl\text{-}C_6H_4C(O)O]_2$	$C_6H_5OC(O)CH_3$	353–376	14.43	127.2	112
$[4\text{-}NO_2\text{-}C_6H_4C(O)O]_2$	$C_6H_5OC(O)CH_3$	353–370	14.40	126.8	112
$[C_6H_5CH_2C(O)O]_2$	$C_6H_5CH_2CH_3$	298–313	12.96	92.0	107
$C_6H_5CH_2C(O)OOC(CH_3)_3$	$C_6H_5CH_2CH_3$	358–383	13.33	118.0	107
$[C_6H_5(CH_2)_2C(O)O]_2$	$C_6H_5CH_2CH_3$	347–358	14.80	126.4	107
$C_6H_5(CH_2)_2C(O)OOC(CH_3)_3$	$C_6H_5CH_2CH_3$	368–386	15.83	147.3	107
$C_6H_5(CH_2)_3C(O)OOC(CH_3)_3$	$C_6H_5CH_2CH_3$	368–383	15.69	145.2	107
$(CH_3)_2CHCH_2C(O)OOC(CH_3)_3$	$C_6H_5CH_2CH_3$	358–383	18.42	164.0	107
$(CH_3)_2CHC(O)OOC(CH_3)_3$	$C_6H_5CH_2CH_3$	358–383	13.43	123.8	107

Table 2.9
Rate constants of monomolecular decay of hydroperoxides in gas phase and aromatic solvents

ROOH	Solvent	T/ K	log (A / s^{-1})	E/ kJ mol^{-1}	Ref.
CH_3OOH	Gas phase	565–651	14.9	180	114
CH_3CH_2OOH	Gas phase	553–653	15.35	180	114
$(CH_3)_2CHOOH$	Gas phase	553–653	15.5	180	114
$(CH_3)_3COOH$	Gas phase	553–653	15.6	180	114
$(CH_3)_3COOH$	C_6H_6	427–447	16.03	170.5	115
$C_6H_5CH_2C(CH_3)_2OOH$	C_6H_6	417–449	10.85	125	115
$C_6H_5CH(OOH)CH(CH_3)_2$	C_6H_6	407–447	10.62	122	115
~$CH(OOH)CH_2$~	C_6H_5Cl	383–403	14.2	146	116

Table 2.10
Rate constants and activation energies of free radical formation by reaction ROOH + Y → free radicals

ROOH	Y	Solvent	T/ K	k/ l mol^{-1} s^{-1} or log k = $A - E/\theta$	Ref.
HOOH	HOOH	*cyclo*-$C_6H_{11}OH$	393–413	$9.84 - 122/\theta$	117
$(CH_3)_3COOH$	ROOH	C_7H_{16}	363	6.3×10^{-7}	118
$(CH_3)_2C(OOH)CH_2CH_3$	ROOH	2-Methylbutane	333–363	$7.80 - 100/\theta$	126
CH_2=$CHCH(CH_3)OOH$	ROOH	CH_2=CHC_2H_5	338–353	$11.70 - 109/\theta$	119
CH_3CH=$CHCH_2OOH$	ROOH	CH_3CH=$CHCH_3$	338–353	$12.70 - 117/\theta$	119
cyclo-[$(CH_2)_2CH(CH_3)(CH_2)_2C(CH_3)OOH$]	ROOH	C_6H_6	333–353	$11.40 - 120/\theta$	120
$C_6H_5C(CH_3)_2OOH$	ROOH	$C_6H_5CH(CH_3)_2$	333–368	$4.86 - 81/\theta$	121
$C_6H_5CH(OOH)CH$=CH_2	ROOH	$C_6H_5CH_2CH$=CH_2	388	2.8×10^{-7}	122
CH_2=$CHCH(OOH)(CH_2)_4CH_3$	ROOH	Octene-1	388	2.9×10^{-7}	122
$C_8H_{17}CHCHCH(O_2H)(CH_2)_6CO_2CH_3$	ROOH	$C_{17}H_{33}COOCH_3$	388	4.5×10^{-7}	122
cyclo-[CH=$CHCH(OOH)(CH_2)_3$]	ROOH	*cyclo*-C_6H_{10}	388	5.4×10^{-7}	122

ROOH	Y	Solvent	T/ K	k/ l mol^{-1} s^{-1} or log k = $A - E/\theta$	Ref.
cyclo-[$CH{=}CHC(CH_3)(OOH)(CH_2)_3$]	ROOH	*cyclo*-$C_6H_9CH_3$	388	1.1×10^{-6}	122
cyclo-[$CHCHC(CH_3)(O_2H)(CH_2)_2C(CH_3)_2$]	ROOH	*cyclo*-$C_6H_7(CH_3)_3$	388	3.2×10^{-6}	122
$C_2H_5CH{=}C(CH_3)CH(OOH)C_2H_5$	ROOH	4-Methylheptene-3	388	1.7×10^{-6}	122
$(CH_3)_2C{=}CHCH(OOH)(CH_2)_3CH_3$	ROOH	2-Methyloctene-2	388	5.0×10^{-7}	122
$C_6H_5C(CH_3)_2OOH$	$C_6H_5CH(CH_3)_2$	C_6H_5Cl	373–393	$7.70 - 109/\theta$	123
$C_6H_5C(CH_3)_2OOH$	$C_6H_5CH(CH_3)_2$	$C_6H_5CH(CH_3)_2$	333–368	$8.00 - 110/\theta$	121
$(CH_3)_3COOH$	$C_6H_5CH{=}CH_2$	C_6H_5Cl	328–363	$4.08 - 72/\theta$	124
$C_6H_5(CH_3)_2COOH$	$CH_2{=}CHCOOCH_3$	—	—	$5.88 - 83/\theta$	125
$C_6H_5(CH_3)_2COOH$	$CH_2{=}CHCN$	—	323	5.4×10^{-8}	125
$C_6H_5(CH_3)_2COOH$	$CH_2{=}C(CH_3)COOCH_3$	—	323–363	$2.74 - 68/\theta$	125
$C_6H_5(CH_3)_2COOH$	$CH_2{=}C(CH_3)COOC_4H_9$	—	323	1.9×10^{-8}	125
$C_6H_5(CH_3)_2COOH$	$CH_2{=}C(C_2H_5)COOCH_3$	—	323	3.2×10^{-8}	125
$C_6H_5(CH_3)_2COOH$	$CH_2{=}C(CH_3)C_6H_5$	C_6H_6	363	7.1×10^{-7}	125
$(CH_3)_2C(OOH)CH_2CH_3$	$(CH_3)_2C(OH)CH_2CH_3$	Isooctane	333–396	$8.50 - 93/\theta$	126
cyclo-$C_6H_{11}OOH$	*cyclo*-$C_6H_{11}OH$	C_6H_5Cl	353–393	$8.64 - 92/\theta$	127
$(CH_3)_3COOH$	*cyclo*-$C_6H_{10}O$	C_6H_5Cl	283–399	$3.40 - 63/\theta$	128
$C_6H_5(CH_3)_2COOH$	*cyclo*-$C_6H_{10}O$	C_6H_5Cl	393	5.7×10^{-6}	129
HOOH	*cyclo*-$C_6H_{10}O$	*cyclo*-$C_6H_{11}OH$	393–413	$-0.32 - 40/\theta$	130
$C_6H_5(CH_3)_2COOH$	$C_6H_5C(O)OH$	$C_6H_5CH(CH_3)_2$	363–383	$9.85 - 86/\theta$	131
$(CH_3)_3COOH$	C_5H_5N	C_6H_5Cl	333–368	$3.54 - 53/\theta$	132
HOOH	CH_3COCH_3	$(CH_3)_2CHOH$	391	1.2×10^{-6}	133

Table 2.11
Rate constants and activation energies of reaction $RH + O_2 \rightarrow R^{\bullet} + HO_2^{\bullet}$

RH	Solvent	T/ K	k/ l mol^{-1} s^{-1} or $\log k = A - E/\theta$	Ref.
$(CH_3)_2CHCH_2CH_3$	C_6H_6	410–439	12.2 – 159/θ	134
$CH_3(CH_2)_5CH_3$	C_6H_6	397–434	14.5 – 181/θ	135
$(CH_3)_3CCH_2CH(CH_3)_2$	C_6H_6	400–466	12.0 – 159/θ	134
cyclo-C_6H_{12}	C_6H_5Cl	383–413	12.9 – 167/θ	136
$C_6H_5CH_3$	C_6H_6	378–418	9.8 – 134/θ	135
2-$CH_3C_6H_4CH_3$	C_6H_5Cl	383–418	9.5 – 130/θ	137
3-$CH_3C_6H_4CH_3$	C_6H_5Cl	383–418	10.5 – 135/θ	137
4-$CH_3C_6H_4CH_3$	C_6H_5Cl	383–418	9.5 – 127/θ	137
$C_6H_5CH(CH_3)_2$	C_6H_6	353–413	7.0 – 113/θ	138
$C_6H_5CH{=}CH_2$	C_6H_5Cl	378–398	11.6 – 125/θ	139
$C_6H_5(CH_3)C{=}CH_2$	C_6H_5Cl	—	8.4 – 105/θ	140
$CH_3(CH_2)_3OCOCH{=}CH_2$	C_6H_5Cl	—	6.0 – 88.6/θ	140
$CH_3(CH_2)_3OCO(CH_3)C{=}CH_2$	C_6H_5Cl	—	6.7 – 91.5/θ	140
$(CH_3)_3COCOCH{=}CH_2$	C_6H_5Cl	353	7.08×10^{-8}	140
$CH_3OCO(CH_3)C{=}CH_2$	C_6H_5Cl	353	1.29×10^{-7}	140
$CH_3OCO(C_2H_5)C{=}CH_2$	C_6H_5Cl	353	1.20×10^{-7}	140
$NH_2COCH{=}CH_2$	C_6H_5Cl	353	1.51×10^{-6}	140
cyclo-$C_6H_{10}O$	C_6H_5Cl	376–413	9.0 – 110/θ	141
$CH_3CH_2CH_2CHO$	C_6H_5Cl	283–303	4.6 – 117/θ	142
$(CH_3)_2CHCHO$	C_6H_5Cl	283–303	8.1 – 92/θ	143
$CH_3(CH_2)_8CHO$	$C_{10}H_{22}$	278–299	5.5 – 65/θ	144
$CH_2{=}CHCHO$	C_6H_6	413	6.31×10^{-2}	145
$CH_3CH{=}CHCHO$	C_6H_5Cl	288–308	15.1 – 124/θ	146
$CH_2{=}C(C_2H_5)CHO$	C_6H_5Cl	288–308	13.4 – 96/θ	147
$CH_2{=}C(C_4H_9)CHO$	C_6H_5Cl	288–308	10.0 – 94/θ	148
$CH_3CH{=}CHCHO$	C_6H_5Cl	273–298	13.9 – 119/θ	148

RH	Solvent	T/ K	k/ l mol^{-1} s^{-1} or log $k = A - E/\theta$	Ref.
$CH_3(CH_2)_3COOH$	2-ClC_6H_4Cl	398–413	11.9 – 143/θ	149
$CH_3OCO(CH_2)_8COOCH_3$	C_6H_5Cl	413–443	13.9 – 164/θ	150
$CH_3CH_2OCO(CH_2)_8COOCH_2CH_3$	C_6H_5Cl	413–443	11.7 – 142/θ	150
$(CH_3)_2CHOCO(CH_2)_8COOCH(CH_3)_2$	C_6H_5Cl	413–443	12.0 – 145/θ	150
$CH_3CH_2C(O)OCH_2CH_3$	C_6H_5Cl	343–363	13.8 – 134/θ	150
$(CH_3CH_2C(O)OCH_2)_4C$	$(CH_3CH_2COOCH_2)_4C$	423	3.39×10^{-6}	151
$(CH_3C(O)OCH_2)_4C$	$(CH_3COOCH_2)_4C$	423	9.33×10^{-7}	151
$(CH_3CH_2C(O)OCH_2)_3CCH_2CH_3$	$(CH_3CH_2COOCH_2)_3CC_2H_5$	423	5.25×10^{-7}	151
$(CH_3CH_2C(O)OCH_2)_2C(CH_3)_2$	$(CH_3CH_2COOCH_2)_2C(CH_3)_2$	423	2.19×10^{-7}	151
$(CH_2{=}C(CH_3)C(O)OCH_2)_4C$	C_6H_5Cl	343–393	18.2 – 181/θ	152

Table 2.12
Rate constants and activation energies of trimolecular reaction 2RH + O_2 → free radicals

RH	Solvent	T/ K	k/ l mol^{-2} s^{-2} or log $k = A - E/\theta$	Ref.
Tetraline	$C_{10}H_{22}$	403–423	3.54 – 86.5/θ	153
Indene	$C_{10}H_{22}$	345–365	3.59 – 78.5/θ	154
cyclo-[$CH{=}CHCH{=}CH(CH_2)_2$]	C_6H_5Cl	313–348	6.85 – 74.5/θ	155
cyclo-[$(CH_3CO)C{=}CH(CH_2)_4$]	C_6H_5Cl	353–393	9.23 – 114/θ	149
C_6H_5CHO	CH_3COOH	303	1.05×10^{-4}	156
C_6H_5CHO	CCl_4	300	5.37×10^{-5}	157
$CH_3CH{=}CHCHO$	$C_{10}H_{22}$	278–298	13.64 – 105/θ	158
$CH_2{=}C(CH_3)CHO$	C_6H_5Cl	288–308	8.10 – 88/θ	146
$C_6H_5CH{=}CH_2$	$HC(O)N(CH_3)_2$	343	2.40×10^{-8}	159
$C_6H_5(CH_3)C{=}CH_2$	C_6H_5CN	343	3.72×10^{-8}	140
$CH_3(CH_2)_3OCO(CH_3)C{=}CH_2$	$HC(O)N(CH_3)_2$	363	6.03×10^{-8}	140
2-$CH_2{=}CHC_5H_4N$	$HC(O)N(CH_3)_2$	343	1.52×10^{-6}	140

RH	Solvent	T/ K	k/ l mol^{-2} s^{-2} or $\log k = A - E/\theta$	Ref.
4- CH_2=CHC_5H_4N	HC(O)N$(CH_3)_2$	343	1.29×10^{-6}	140
5-(2-$CH_3C_5H_3N$)CH=CH_2	HC(O)N$(CH_3)_2$	343	2.82×10^{-6}	140
$C_8H_{17}CH$=CH$(CH_2)_7COOCH_3$	C_6H_5Cl	313–333	12.30 – 142/θ	159
$CH_3(CH_2)_3(CH_2CH$=CH$)_2(CH_2)_7COOCH_3$	C_6H_5Cl	313–333	6.30 – 93/θ	159
C_2H_5(CH=CH$CH_2)_3(CH_2)_6COOCH_3$	C_6H_5Cl	313–333	3.80 – 76.5/θ	159
Cholesterylpelargonate	C_6H_5Cl	364–388	5.44 – 73/θ	160
cyclo-[OCH=CHCH=C($COOCH_3$)]	C_6H_5Cl	303–363	1.97 – 54/θ	161
cyclo-[OCH=CHCH=C($COOC_4H_9$)]	C_6H_5Cl	313–363	2.15 – 62/θ	161
cyclo-[OCH=CHCH=C(CH=CH$COOC_2H_5$)]	C_6H_5Cl	313–363	0.25 – 52/θ	161
cyclo-[OCH(C_3H_7)O$(CH_2)_3$]	C_6H_5Cl	383–403	2.59 – 74.5/θ	162
cyclo-[$OCH_2OC(CH_3)_2(CH_3)_2$]	C_6H_5Cl	343–403	0.47 – 60/θ	162
cyclo-[$OCH_2OCH(C_6H_5)(CH_2)_2$]	C_6H_5Cl	373–393	2.77 – 72.5/θ	162
cyclo-[OCH(CH=CH_2)O$(CH_2)_3$]	C_6H_5Cl	373–393	5.07 – 84/θ	162
$(4\text{-}NCC_6H_4)_2CH_2$	$C_6H_3Cl_3$	373–413	3.68 – 63/θ	163
$[CH_3(CH_2)_3O]_2CH_2$	C_6H_5Cl	383–413	1.54 – 73/θ	164
$[CH_3(CH_2)_3O]_2CHCH_3$	C_6H_5Cl	373–413	3.72 – 83.6/θ	164

REFERENCES

1. **Mill, T., Montorsi, G.,** The liquid-phase oxidation of 2,4-dimethylpentane, *Int. J. Chem. Kinet.*, 5, 119, 1973.
2. **Van Sickle, D. E., Mill, T., Mayo, F. R., Richardson, H., Gould, C. W.,** Intramolecular propagation in the oxidation of *n*-alkanes. Autoxidation of *n*-pentane and *n*-octane, *J. Org. Chem.*, 38, 4435, 1973.
3. **Mill, T., Hendry, D. G.,** *Kinetic and mechanism of free radical oxidation of alkans and olefines in the liquid phase, in Comprehensive Chemical Kinetics*, Ed. Bamford C.H., Tipper, C. F. H., Elsevier Publ. Co., Amsterdam, V.16, 1980, p. 44.
4. **Van Sickle, D. E.,** Oxidation of 2,4,6-trimethylheptane, *J. Org. Chem.*, 37, 755, 1972.
5. **Demidov, I. N., Solyanikov, V. M.,** The selectivity of oxidation of paraffines and mechanism of chain propagation, *Neftekhimiya*, 26, 406, 1986.
6. **Jensen, R. K., Korcek, S., Mahoney, L.R., Zinbo, M.,** Liquid-phase autoxidation of organic compounds at elevated temperatures. 2. Kinetics and mechanisms of the formation of cleavage products in n-hexadecane autoxidation, *J. Am. Chem. Soc.*, 103, 1742, 1981.
7. **Howard, J. A., Ingold, K. U.,** Absolute rate constants for hydrocarbon autoxidation. 18. Oxidation of some acyclic ethers, *Can. J. Chem.*, 48, 873, 1970.

8. **Opeida, I. A., Timokhin, V. I., Galaz, V. E.**, Isomerisation of peroxyl radical of diisopropylketone, *Theor. Experim. Khim.*, 14, 554, 1978.

9. **Sharafutdinova, Z. F., Martemyanov, V. S., Borisov, I. M.**, Initiated oxidation of propionates of polyatomic alcohols, *Kinet. Katal.*, 29, 553, 1988.

10. **Bothe, E., Behrens, G., Schulte-Frolinde, D.**, Mechanism of the first order decay of 2-hydroxypropyl-2-peroxyl radicals and of O_2^- formation in aqueous solution, *Z. Naturforsch.* B. 32, 886, 1977.

11. **Ladygin, B. Ya., Revin, A. A.**, Kinetics of reaction of radicals generated by radiolysis of cycloalkanols in the liquid-phase at the presence of oxygen, *Izv. Akad. Nauk SSSR. Ser. Khim.*, 1985, 282.

12. **Bothe, E., Schulte-Frolinde, D., von Sonntag, G.**, Radiation chemistry of carbohydrates. Part 16. Kinetics of $HO_2^•$ elimination from peroxyl radicals derived from glucose and polyhydric alcohols, *J. Chem. Soc. Perkin Trans.* 2, 1978, 416.

13. **Koelewijn, P.**, Epoxidation of olefines by alkylperoxy radicals, *Rec. Trav. Chim. Pay-Bas*, 91, 759, 1972.

14. **Mayo, F. R., Syz, M. G., Mill, T., Castleman, J. K.**, Co-oxidation of hydrocarbons, *Adv. Chem. Ser.*, 75, 38, 1968.

15. **Machtin, V. A.**, *Reactions of peroxyl radicals in oxidizing vinyl monomers and reactivity of double bonds*, Cand. Sci. (Chem.) Thesis Dissertation, Inst. Chem. Phys., Chernogolovka, 1984, 1-30 (in Russian).

16. **Chevriau, C., Naffa, P., Balaceanu, J. C.**, Oxidation competitive des hydrocarbures en phase liquide. Reactivite des radicaux polyperoxidiques issus d'un hydrocarbure vinylique: cas du styrene, *Bull. Soc. Chim. France 1*, 1964, 3002.

17. **Howard, J. A., Ingold, K. U.**, Absolute rate constants for hydrocarbon autooxidation. 1. Styrene, *Can. J. Chem.*, 43, 2729, 1965.

18. **Howard, J. A., Ingold, K. U.**, Absolute rate constants for hydrocarbon autoxidation. 2. Deutero styrenes and ring substituted styrenes, *Can. J. Chem.*, 43, 2737, 1965.

19. **Howard, J. A., Ingold, K. U.**, Absolute rate constants for hydrocarbon autoxidation. 3. α-Methylstyrene, β-methylstyrene and indene, *Can. J. Chem.*, 43, 2737, 1965.

20. **Howard, J. A.**, Absolute rate constants for hydrocarbon autoxidation. 22. The autoxidation of some vinyl compounds, *Can. J. Chem.*, 50, 2298, 1972.

21. **Opeida, I. A., Nechitaylo, L. G.**, About addition of alkylaromatic peroxyl radicals to aromatic molecules, *Kinet. Katal.*, 19, 1581, 1978.

22. **Howard, J. A., Ingold, K. U.**, Absolute rate constants for hydrocarbon autoxidation. 6. Alkyl aromatic and olefinic hydrocarbons, *Can. J. Chem.*, 45, 793, 1967.

23. **Rubailo, V. L., Gagarina, A. B., Emanuel, N. M.**, Rate constants of chain propagation and termination in liquid-phase oxidation of 1-phenylcyclohexene, *Dokl. Akad. Nauk SSSR*, 224, 883, 1975.

24. **Pliss, E. M., Alexandrov, A. L., Mogilevich M. M.**, Rate constants of peroxy radicals in liquid phase oxidation reactions of methacrylic and acrylic esters, *Izv. Akad. Nauk SSSR, Ser. Khim.*, 1975, 1971.

25. **Machtin, V. A., Pliss, E. M., Denisov, E. T.**, Reactions of peroxyl radicals with acrylic and methacrylic esters of polyatomic alcohols, *Izv. Akad. Nauk SSSR, Ser. Khim.*, 746, 1981.

26. **Thomas, J. K.**, Pulse radiolysis of aqueous solutions of methyl iodide and methyl bromide, *J. Phys. Chem.*, 71, 1919, 1967.

27. **Hickel, B.**, Absorption spectra and kinetics of methyl and ethyl radicals in water, *J. Phys., Chem.*, 79, 1054, 1975.

28. **Fessenden, R. W., Carton, P. M., Shimamori, H.**, Measurement of the dipole moments of excited states and photochemical transients by microwave dielectric absorption, *J. Phys. Chem.*, 86, 3803, 1982.

29. **Brede, O., Herman, R., Mehnert, R.**, Primary processes of stabilizer action in radiation induced alkane oxidation, *J. Chem. Soc. Faraday Trans. 1*, 83, 2365, 1987.

30. **Hasegawa K., Patterson L. K.**, Pulse radiolysis studies in model lipid systems: formation and behavior of peroxy radicals in fatty acids, *Photochem. Photoliol*, 28, 817, 1978.

31. **Mailard, B., Ingold K. U., Scaiano, J. C.**, Rate constants for the reactions of free radicals with oxygen in solution, *J. Am. Chem. Soc.*, 105, 5095, 1983.

32. **Schulte-Frohlinde, D., Anker, R., Bothe, E.,** *Hydroxyl radical induced oxidation of acetylene in oxygenated aqueous solution., Oxygen and oxy-radicals in chemistry and biology*, Ed. Rogers, M. A. J. and Powers, E. L., Academic Press, New York, N.Y., 1981, 61-67.

33. **Neta, P., Huie, R. E., Mosseri, S., Shastry L. V., Mettal, J. P., Maruthamuthu, P., Steenken, S.,** Rate constants of substituted methylperoxyl radicals by ascorbat ions and N,N,N',N'-tetramethyl-*p*-phenylendiamine, *J. Phys. Chem.*, 93, 4099, 1989.

34. **Alexandrov, A. L., Pliss, E. M., Shuvalov, V. F.,** Rate constants of reaction of alkyl radicals with oxygen and stable nitroxyl radicals, *Izv. AN SSSR. Ser. Khim.*, 1979, 2446.

35. **Zador, E., Warman, J., Hummel, A.,** Formation of the trityl cation in pulse irradiated solutions of triphenylmethylion in pulse irradiated solutions of triphenylmethyl chloride in cyclohexane. *J. Chem. Soc. Faraday Trans. 1*, 75, 914, 1979.

36. **Bennett, J. E., Brunton, G., Smith, J. R. L., Salmon, T. M. F., Waddington, D. J.,** Reactions of alkylperoxyl radicals in solution. P.1. A kinetic study of self-reactions of 2-propylperoxyl radicals between 135-300 K, *J. Chem. Soc. Faraday Trans.1*, 83, 2421, 1987.

37. **Howard, J. A., Ingold, K. U.,** Rate constants for the self reactions of *n*-and *sec*-butylperoxy radicals and cyclohexylperoxy radicals. The deuterium isotope effect in the termination of secondary peroxy radicals, *J. Am. Chem. Soc.*, 90, 1058, 1968.

38. **Howard, J. A., Bennett, J. E.,** The self-reaction of *sec*-alkylperoxy radicals. A kinetic electron spin resonance study, *Can. J. Chem.*, 50, 2374, 1972.

39. **Maguire, W. J., Pink, R. C.,** Studies by electrone spin resonance of radicals produced in the photolysis of alkyl peroxides and hydroperoxides, *Trans. Faraday Soc.*, 64, 1097, 1967.

40. **Bennett, J. E., Brown, D. M., Mile, B.,** Studies by electron spin resonance of the reactions of alkylperoxy radicals, *Trans. Faraday Soc.*, 66, 386, 1970.

41. **Maslennikov, S. I., Nikolaev, A. I., Komissarov, V. D.,** The kinetics of disproportionation of secondary peroxyalkyl radicals of *n*-alkanes, *Kinet. Katal.*, 20, 326, 1979.

42. **Faucitano, A., Buttafava, A., Martinotti, F., Comincoli, V., Bortolus, P.,** Decay constants of peroxy radicals from polypropylene and polypropylene model compounds, *Polymer Photochem.*, 7, 491, 1986.

43. **Furimsky, E., Howard, J. A., Selwyn, J.,** Absolute rate constants for hydrocarbon autoxidation. 28. A low temperature kinetic electron spin resonance study of the selfreactions of isopropylperoxy and related secondary alkylperoxy radicals in solution, *Can. J. Chem.*, 58, 677, 1980.

44. **Maslennikov, S. I., Galimova, I. G., Komissarov, V. D.,** Disproportionation rate and products of cyclohexylperoxy radicals, *Izv. Akad. Nauk SSSR, Ser. Khim.*, 1979, 631.

45. **Smaller, B., Retko, J. R., Avery, E. C.,** Electron paramagnetic resonance studies of transient free radicals produced by pulse radiolysis, *J. Chem. Phys.*, 48, 5174, 1968.

46. **Ladygin, B. Ya., Zimina, G. M., Vannikov, A. V.,** Kinetics of reactions of peroxy radicals formed by the electron irradiation of normal and cyclic alkanes, *Khim. Vys. Energ.*, 1984

47. **Zaikov, G. E., Maizus, Z. K., Emanuel, N. M .,** Solvation influence on reactivity of peroxyl. radicals with double bond, *Izv. Akad. Nauk SSSR, Ser. Khim.*, 1968, 2265.

48. **Ickovich, V. A., Potekhin, V. M., Pritzkow, W., Proskuryakov, V. A., Schnurpfeil, D.,** *Autoxidation von Rohlenwassertoffen*, VEB Deutsch. Verlag Gerundstoffindustrie, Berlin, 1981, 183.

49. **Howard, J. A., Ingold, K. U.,** Absolute rate constants for hydrocarbon autoxidation.4. Tetraline, cyclohexene, diphenyl methane, ethylbenzene and allylbenzene. *Can. J. Chem.*, 44, 1119, 1966.

50. **McCarthy, R. L., MacLachlan, A.,** Kinetics of some radiation induced reactions, *Trans. Faraday Soc.*, 57, 1107, 1961.

51. **Rousseau-Richard, C., Richard, C., Martin, R.,** Etude cinetique de l'influence compliexe, proou antioxidante, de le vives phenoliques sur l'oxidation induite d'un substrat polyinsature, *J. Chim. Phys.Chim. Biol.*, 85, 175, 1988.

52. **Barclay, L. R. C., Baskin, K. A., Locke, S. J., Schaefer, T. D.,** Benzophenone-photosensitised autoxidation of linoleate in solution and sodium dodecyl sulfate micells, *Can. J. Chem.*, 65, 2529, 1987.

53. **Bateman, L., Gee, G., Morris, A. L., Watson, W. F.,** The velocity coefficients of the chain propagation and termination reactions in olefin oxidation in liquid systems, *Disc. Faraday Soc.*, 1951, 250.

54. **Vardanyan, R. L., Safiullin, R. L., Komissarov, V. D.,** Absolute values of rate constants of disproportionation of peroxy radicals formed from cholesterol, *Kinet. Katal.* 26, 1140, 1985.
55. **Kenisberg, T. P., Ariko, N. G., Mitskevich, N. I.,** Estimation of rate constants of recombination of peroxyl radicals of *p*-xylene, *Dokl. Akad. Nauk BSSR,* 24, 817, 1980.
56. **Robb, J. C., Shanin, M.,** A thermocouple method of studying oxidation reactions. 2. Photosensitized oxidation of 1-methyl, -4-methyl and 4,5-dimethyl-cyclohexene, *Trans. Faraday Soc.*, 55, 1753, 1959.
57. **Zaikov, G. E., Howard, J. A., Ingold, K. U.,** Absolute rate constants for hydrocarbon oxidation. 13. Aldehydes: photo-oxidation, co-oxidation, and inhibition, *Can. J. Chem.*, 47, 3017, 1969.
58. **Nikolayevskii, A. N., Koloyerova, V. G., Kucher, R. V.,** Effectivity of inhibition of ethylbenzene liquid phase oxidation with dioxybenzenes in acetic acid medium, *Neftekhimiya*, 16, 752, 1976.
59. **Belyakov, V. A., Lauterbach, G., Pritzkow, W., Voerckel, V.,** Kinetics and regioselectivity of the autoxidation of alkylaromatic hydrocarbons, *J. Prakt. Chem. (Leipzig)*, 340, 475, 1992.
60. **Opeida, I. A., Matvienko, A. G., Yefimova, I. V., Kachurin I. O.,** On rate constants of chain propagation and termination at oxidation of substituted ethylbenzenes and isomeric ethylpyridines, *Zh. Org. Khim.*, 24, 572, 1988.
61. **Opeida, I. A., Matvienko, A. G., Yefimova, I. V.,** On reactivity of polyethylbenzenes in reaction with peroxy radicals, *Kinet. Katal.*, 28, 1341, 1987.
62. **Gaponova, I. S., Fedofova, T. V., Tsepalov V. F.,** The study of recombination of cumylperoxyl radicals in liquid and supercooled solutions, *Kinet. Katal.*, 12, 1137, 1971.
63. **Howard, J. A., Ingold, K. U., Symonds, M.,** Absolute rate constants for hydrocarbon oxidation. 8. The reactions of cumylperoxy radicals, *Can. J. Chem.*, 46, 1017, 1968.
64. **Howard, J. A., Ingold, K. U.,** Absolute rate constants for hydrocarbon oxidation. 11. The reactions of tertiary peroxy radicals, *Can. J. Chem.*, 46, 2655, 1968.
65. **Howard, J. A., Robb, J. C.,** Thermocouple method for studying oxidation reactions. 3. The photosensitized oxidation of cumene, cyclohexene and *p*-cumene, *Trans. Faraday Soc.*, 59, 487, 1963.
66. **Gerasimova, S. A., Kachurin, I. O., Matvienko, A. G., Opeida, I. A.,** Reaction of chain termination in liquid-phase oxidation of polyisopropylbenzenes, *Neftekhimiya*, 30, 476, 1990.
67. **Opeida, I. A., Yefimova, I. V., Matvienko, A. G., Dmitruk, A. F., Zarechnaya, O. M.,** Structure and reactivity of some ethylarenes in reaction with peroxyl radicals, *Kinet. Katal.*, 31, 1342, 1990.
68. **Bennett, J. E.,** Kinetic electron paramagnetic resonance study of the reactions of tert-butylperoxyl radicals in aqueous solution, *J. Chem. Soc. Faraday Trans. 2*, 86, 3247, 1990.
69. **Gasborne, P., Seree de Roch, I.,** Etude cinetique de l'oxydation des ethers en phase liquide, *Bull. Soc. Chim. France*, 1967, 2260.
70. **Opeida, I. A.,** *Cooxidation of alkylaromatic hydrocarbons in the liquid phase*, Doct. Sci. (Chem) Dissertation, Inst. Chem. Phys., Chernogolovka, 1981, 1-336 (in Russian).
71. **Rafikova, V. S., Brin, E. F., Skibida, I. P.,** Reactivity of alkylaromatic and alkylperoxyl radicals in reactions with hydrocarbons, *Kinet. Katal.*, 12, 1374, 1971.
72. **Niki, E., Kamia, J., Ohta, N.,** The cooxidation of hydrocarbons in liquid phase, *Bull. Chem. Soc. Japan*, 42, 512, 1969.
73. **Parlant, C.,** Cooxidation de hydrocarbures, *Rev. Inst. France Petrol.*, 19, 1, 1964.
74. **Timokhin, V. I., Opeida, I. A., Kucher, R. V.,** Cooxidation of alkylaromatic hydrocarbons, *Neftekhimiya*, 17, 555, 1977.
75. **Kharlampidi, Kh. E., Nigmatullina, F. I., Batyrshin, F. I., Lebedeva, N. M.,** Effect of small additives of styrene on kinetics of liquid-phase oxidation of ethylbenzene, *Neftekhimiya*, 24, 676, 1986.
76. **Sajus, L.,** Kinetic data on the radical oxidation of petrochemical compounds, *Adv. Chem. Ser.*, 75(1), 59, 1968.
77. **Machtin, V. A.,** *Reactions of peroxyl radicals in oxidizing vinyl monomers and reactivity of double bonds*, Cand. Sci. (Chem) Dissertation, Inst. Chem. Phys. Chernogolovka, 1984, P.1-130 (in Russian).
78. **Prochazka, M., Ryba D., Lim D.,** Azo-compounds. II. The kinetics of thermal decomposition 2,2'-azoizobutane 2,2-dimethyl-2,2'-azopentane, and-1,1'-azoadamantane, *Collect. Chem. Comm.* 33, 3387, 1968.

79. **Bandlish, B. K., Garner, A. W., Hodges, M. L., Timberlake, J. W.**, Substituent effects in radical reactions., III. Thermolysis of substituted phenylazomethanes, 3,5-diphenyl-1-pyrazolines, and azopropanes, *J. Am. Chem. Soc.*, 97, 5856, 1975.
80. **Overberger, C. G., DiGiulio, A. U.**, Azo Compounds. XXIX. Decomposition study of a-alkyl- and a,a-dialkylbenzylalkanes, *J. Am. Chem. Soc.* 81, 2154, 1959.
81. **Ohno, A., Ohnishi, Y.**, Resonance participation of sulfur and oxygen in radicals decomposition of azobis(2-propane) derivatives, *Tetrahedron Lett.* 1969, 4405.
82. **Cohen, S. G., Wang, Ch. -H.**, Phenyl-azo-diphenylmethane and the decomposition of azo compounds, *J. Am. Chem. Soc.* 77, 3628, 1955.
83. **Cohen, S. G., Cohen, F., Wang, Ch. -H.**, Comparison of 9-phenylfluorenyl and triphylmethyl in the decomposition of compounds, *J. Org. Chem.* 28, 1479, 1963.
84. **Cohen, S. G., Wang, Ch. -H.**, Azo-bis-diphenylmethane and the decomposition of aliphatic azo compounds. The diphenylme thyl radical, *J. Am. Chem. Soc.* 77, 2457, 1955.
85. **Shelton, J. R., Liang, Ch. -K., Kovacic, P.**, Chemistry of diarylazoalkanes. II. Effect of *para*-substituents on the thermal decomposition of azocumenes, *J. Am. Chem. Soc.* 90, 354, 1968.
86. **Overberger, C. G., Lombardino, J. B.**, Azo compounds. A seven-membered cyclic azo compound, *J. Am. Chem. Soc.*, 80, 2317, 1958.
87. **Overberger, C. G., Tashlich, J.**, Azo compounds. An eight-membered cyclic azo compound, *J. Am. Chem. Soc.*, 81, 217, 1959.
88. **Martin, J. C., Timberlake, J. W.**, Kinetic studies of reactions to cyclopropylcarbinyl radicals. Cyclopropylsubstituted azomethanes and hexacyclopropylethane, *J. Am. Chem. Soc.* 92, 978, 1970.
89. **Ziegler, K., Deparade, W., Meye W.**, Zur Kennthis des "dreiwertigen" Kohlenstoffs. XXIII: Der Zerfall des Azo-isobuttersaurenitrils und vervandter Substanzen, *Lieb. Ann.*, 567, 141,1950.
90. **Lim, D.**, On the kinetics of the decomposition of azo compounds. I. The decomposition of azo-bis-nitriles and esters of azo-bis-isobutyricacid, *Collect. Czech. Chem. Comm.*, 33, 1122, 1968.
91. **Overberger, C. G., O'Shaughnessy, M. T., Shalit, H.**, The preparation of some aliphatic azonitriles and their decomposition in solution, *J. Am. Chem. Soc.*, 71, 2661, 1949.
92. **Overberger, C. G., Biletch, H., Fenistone, A. B., Lilker, J., Herbert, J.**, Azo bisnitriles. The decomposition of azocompounds derived from cyclyalkanones. An accurate measure of difference in ring strain, *J. Am. Chem. Soc.*, 75, 2078, 1953.
93. **Martin, J. C., Schults, J. E., Timberlake, J. W.**, Cyclopropylcarbinyl free radicals. Hexacyclopropylethane, *Tetrahedron Lett.*, 1967, 4629.
94. **Neumann, W. P., Ling, H.**, Zum induzierten radikalischen Zerfall von Azoverbindungen RO-N=N-OR. *Chem. Ber.*, 101, 2837, 1968.
95. **Cavell, E. A. S., Meeks, A. C.**, Temperature dependence of the rate of initiation of polymerization by 4,4'-azo-bis-cyanopentanoic acid, *Makromol. Chem.*, 108, 304, 1967.
96. **Vernekar, S. P., Ghatge, N. D., Wadgaoknar, P. P.**, Decomposition rate studies of azobisnitriles containing functional groups, *J. Polymer Sci. A-1.*, 1988, 26, 953, 1988.
97. **Zawalski, R .C., Lisiak, M., Kovacic, P., Luedtke A., Timberlake, J. M.**, Radical stabilization synthesis and decomposition of b-ketodiazene, *Tetrahedron Lett.*, 21, 425, 1980.
98. **Pryor, W. A., Huston, D. M., Fiske, T. R., Rickering, T. L., Ciuffarin, E.**, Reaction of radicals. 11. Ethyl peroxide, isopropyl peroxide and *sec*-butyl peroxide, *J. Am. Chem. Soc.*, 86, 4237, 1964.
99. **Pryor, W. A., Kaplan, C. L.**, Reactions of radicals. 10. Butyl peroxide, *J. Am. Chem. Soc.*, 86, 4234, 1964.
100. **Denisov, E. T.**, *Rate Constants of Liquid-phase Reactions*, Plenum Press, N. Y., 1973.
101. **Mashnenko, O. M., Batog, A. E., Mironenko, N. I., Romantsevich, M. K.**, The polymerisation of styrene initiated by alkyl-*tert*-alkylperesters of dimethyleneglicoles, *Vysokomolek. Soed. B.*, 10, 444, 1968.
102. **Hiatt, R. R.**, The decomposition of silyl peroxides, *Can. J. Chem.*, 42, 985, 1964.
103. **Antonovskii, V. L., Bezborodova, L. D.**, The study of initiating effectiveness of organic peroxides by method of inhibitors, *Zh. Fiz. Khim.*, 44, 1224, 1970.
104. **Levy, M., Steinberg, M., Szwarc, M.**, Kinetics of the thermal decomposition of diacetyl peroxide. II. Effects of solvents on the rate of the decomposition, *J. Am. Chem. Soc.*, 76, 5978, 1954.

105. **Antonovskii, V. L., Bezborodova, L. D., Yaselman, M.**, The study of ability organic peroxides as initiators by means inhibitors, *Zh. Fiz. Khim.*, 43, 2286, 1969.

106. **Strain, F., Bissinger, W. E., Dial, W. R., Rudoff, H., DeWitt B. J., Stevens, H. C., Langston, J. H.**, Esters of peroxycarbonic acids, *J. Am. Chem. Soc.* 72, 1254, 1950.

107. **Voloshanovskii, I. S., Ivanchev, S. S.**, Thermal decay of aliphatic peroxides and *tert*-butylperesters, *Zh. Obshch. Khim.*, 44, 892, 1974.

108. **Lamb, R. C., Pacifici, J. G., Ayers, P. W.**, Organic peroxides. IV. Kinetics and products of decompositions of cyclohexaneformyl and isobutyryl peroxides. BDPA as a free-radical scavenger, *J. Am. Chem. Soc.* 87, 3928, 1965.

109. **Kartasheva, Z. S., Kasaikina, O. T.**, Thermal decay of dicyclohexylperoxidicarbonate in different solvents, *Izv. Akad. Nauk SSSR, Ser. Khim.*, 1991, 48.

110. **Pankevich, R. V., Dutka, V. S.**, Kinetics of thermal destruction of diacyl peroxides in the presence of transition metaloxides, *Kinet. Katal.*, 33, 1087, 1992.

111. **Shine, H. J., Waters, J. A., Hoffman, D. M.**,The decomposition of acetyl peroxide in solution. III. Kinetics and use of radical traps, *J. Am. Chem. Soc.*, 85, 3613, 1963.

112. **Blomquist, A.T., Buselli, A.J.**, The decomposition of sym-substituted benzoyl peroxides, *J. Am. Chem. Soc.* 73, 3883, 1951.

113. **Swain, C. G., Stockmayer, W. H., Clarke, J. T.**, Effect of structure on the rate of spontaneous thermal decomposition of substituted benzoyl peroxides, *J. Am. Chem. Soc.* 72, 5426, 1950.

114. **Richardson, W. H., O'Neal, H. E.**,in *Comprehensive Chemical Kinetics*, Ed. Bamford, C.H., Tipper, C.F.H., Elsevier Publ. Co., Amsterdam, 1972, 539.

115. **Hiatt, R. R., Strachan, W. M. J.**, The effect of structure on the thermal stability of hydroperoxides, *J. Org. Chem.*, 28, 1893, 1963.

116. **Zolotova, N. V., Denisov, E. T.**, Decomposition of polyethylene hydroperoxide into free radicals in solution and solid phase, *Vysokomolek. soed. B*, 12, 866, 1970.

117. **Denisov, E. T., Kharitonov, V. V.**, Generation of free radicals by hydrogen peroxide in cyclohexanole, *Kinet. Katal.*, 5, 781, 1964.

118. **Denisov, E. T.**, The role of hydrogen bonding in generation of free radicals from hydroperoxide, *Zh. Fiz. Khim.*, 38, 2085, 1964.

119. **Chauvel, A., Clement, G., Balaceanu, J. C.**, Oxidation des olefines en phase liquide. 2. Oxidation comparees des butenes-1 et 2, *Bull. Soc. Chim. France*, 1963, 2025.

120. **Tobolsky, A. V., Mesrobian, R. E.**, *Organic peroxides*, N. Y., 1954.

121. **Solomko, N. I., Tsepalov, V. F., Yurzhenko, A. I.**, Some peculiarities of homogeneous and emulcian oxydation of cumene. *1.* Homogeneous oxidation of cumene, *Kinet. Katal.*, 9, 766, 1968.

122. **Bateman, L.**, Mechanism of autoxidation of olefines, *Quart. Rev.*, 8, 147, 1954.

123. **Antonovskii, V. L., Denisov, E. T., Solntseva, L. V.**, The study of mechanism of liquid-phase oxidation of cumene with inhibitors. 2. Mechanism of decomposition of hydroperoxide, *Kinet. Katal.*, 6, 815, 1965.

124. **Denisov, E. T., Denisova, L. N.**, Generation of free radicals by reaction of hydroperoxide with double bond of styrene, *Dokl. Akad. Nauk SSSR*, 157, 907, 1964.

125. **Pliss, E. M., Troshin, V. M.**, Mechanism of free radicals generation in oxidizing olefinic compounds, *Neftekhimiya*, 22, 539, 1982.

126. **Degtyareva, T. G., Solyanikov, V. M., Denisov, E. T.**, Mechanism of degenerate chain branching in oxidizing isopentane, *Neftekhimiya*, 12, 854, 1972.

127. **Semenchenko, A. E., Solyanikov, V. M., Denisov, E. T.**, Mechanism of generation of free radicals in oxidizing cyclohexane, *Neftekhimiya*, 10, 864, 1970.

128. **Denisov, E. T.**, Generation of free radicals by reaction of hydroperoxide with cyclohexanone, *Dokl. Akad. Nauk SSSR*, 146, 394, 1962.

129. **Denisov, E. T.**, Generation of free radicals on reaction of hydroperoxide with ketones, *Zh. Fiz. Khim.*, 37, 1896, 1963.

130. **Denisov, E. T., Kharitonov, V. V., Raspopova, E. N.**, Generation of free radicals by reaction of hydrogen peroxide with cyclohexanone, *Kinet. Katal.*, 5, 981, 1964.

131. **Antonovskii, V. L., Denisov, E. T., Solntseva, L. V.**, The study of mechanism of liquid-phase oxidation of cumene by method of inhibitors. 3. Products of reaction and their influence on decomposition of hydroperoxide, *Kinet. Katal.*, 7, 409, 1966.

132. **Zolotova, N. V., Denisov, E. T.**, Generation of free radicals by reaction of hydroperoxide with pyridine, *Izv. Akad. Nauk SSSR, Ser. Khim.*, 1966, 767.

133. **Denisov, E. T., Solyanikov, V. M.**, Mechanism of liquid-phase oxidation of isopropanol, *Neftekhimiya*, 4, 458, 1964.

134. **Degtyareva, T. G., Denisova, L. N., Denisov, E. T.**, Generation of free radicals in the system RH + O_2. 3. 2-Methylbutane, 2,2,4-trimethylpentane, *Kinet. Katal.*, 13, 1400, 1972.

135. **Shafikov, N. Ya., Denisova, L. N., Denisov, E. T.**, Generation of free radicals in the system RH + O_2 5. *n*-Heptane, toluene, cyclohexene, *Kinet. Katal.*, 16, 872, 1975.

136. **Denisova, L. N., Denisov, E. T.**, Generation of free radicals in the system RH + O_2. 2. Cyclohexane, *o*-xylene, cumene, *Kinet. Katal.*, 10, 1244, 1969.

137. **Kutuev, A. A., Terpilovskii, N. I.**, Chain generation at liquid-phase oxidation of xylenes, *Kinet. Katal.*, 16, 372, 1976.

138. **Denisov, E. T., Denisova, L. N.**, Estimation of the R-H Bond energy from the activation energy for the reaction $RH + O_2 \rightarrow R^{\bullet} + HO_2^{\bullet}$, *Int. J. Chem. Kinet.*, 8, 123, 1976.

139. **Denisova, L. N., Denisov, E. T.**, Generation of free radicals by reaction oxygen with double bond of styrene, *Izv. Akad. Nauk SSSR, Ser. Khim.*, 1965, 1702.

140. **Pliss, E. M.**, *Oxidation of vinyl-compounds: Mechanism, elementary acts, reactivity via structures*, Doct. Sci. (Chem.) Thesis Dissertation, Inst. Chem. Phys., Chernogolovka, 1990, p. 12–14 (in Russian).

141. **Butovskaya, G. V., Agabekov, V. E., Dmitrieva, O. P., Mitskevich, N. I.**, Generation of chains at liquid-phase oxidation of cyclohexanone, *Dokl. Akad. Nauk BSSR*, 20, 1103, 1976.

142. **Chernyak, B. I., Andrianova, L. A.**, On the mechanism of chain generation in oxidizing butyronitryl, *Zh. Org. Khim.*, 11, 1800, 1975.

143. **Chernyak, B. I., Andrianova, L. A.**, Kinetics of liquid-phase oxidation of isobutyric aldehyde, *Neftekhimiya*, 14, 97, 1974.

144. **Cooper, H. R., Melville, H. W.**, The kinetics of autoxidation of *n*-decanal. Part 1. The mechanism of reaction, *J. Chem. Soc.*, 1951, 1984.

145. **Hava, T., Ohkatsu, Y., Osa, T.**, Autoxidation of acrolein, *Chem., Lett*, 1973, 1953.

146. **Maaraui, M. A.**, *Synthesis of alkylacroleins and study of its liquid phase oxidation*, Doct. Sci. (Chem.) Thesis Dissertation, Lvov, 1978 (in Russian).

147. **Maaraui, M. A., Tolopko, D. K., Chernyak, B. I.**, Chain generation at liquid-phase oxidation of α-ethylacroleins, *Kinet. Katal.*, 18, 224, 1977.

148. **Maaraui, M. A., Chernyak, B. I., Nikipanchuk, M. V.**, Chain generation at the liquid-phase oxidation of crotonyl aldehyde, *Zh. Org. Khim.*, 16, 1573, 1978.

149. **Butovskaya, G. V., Agabekov, V. E., Mitskevich, N. I.**, Reactions of chain generation and propagation at liquid-phase oxidation of valeric acid, *Dokl. Akad. Nauk BSSR*, 22, 155, 1978.

150. **Agabekov, V. E.**, *The reactions and reactivity of oxygen-containing compounds in free radical reactions of oxidation*, Doct. Sci. (Chem.) Thesis Dissertation, Inst. Chem. Phys., Chernogolovka, 1980 (in Russian).

151. **Pozdeeva, N. N., Denisov, E. T.**, Multidipole interaction in reactions of oxygen with esters of polyatomic alcohols, *Izv. AN SSSR, Ser. Khim.* 1987, 2681.

152. **Pliss, E. M., Troshin, V. M., Denisov, E. T.**, Multidipoles interaction in reaction of oxygen with double bond, *Dokl. Akad. Nauk SSSR*, 264, 368, 1982.

153. **Denisov, E. T.**, Generation of free radicals in the system RH + O_2. 1. Tetraline cyclohexanol, cyclohexanone, *Kinet. Katal.*, 4, 53, 1963.

154. **Carlsson, D. J., Robb, J. C.**, Liquid-phase oxidation of hydrocarbons. Part-4. Indene and tetraline: Occurence and mechanism of the thermal initiation reaction with oxygen, *Trans. Faraday Soc.*, 62, 3403, 1966.

155. **Vardanyan, R. L., Verner, I. G., Denisov, E. T.,** Generation of free radicals in the system RH + O_2. 4. 1,3-Cyclohexadien, *Kinet. Katal.,* 14, 575, 1973.

156. **Boga, E., Marta, F.,** Oxidation of benzaldehyde catalyzed by transition metal ions, *Acta Chim. Acad Sci Hung,* 78, 105, 1973.

157. **Komissarova, I. N.,** *Mechanism of oxidation of aldehydes by ozone oxygen,* Cand. Sci. (Chem.) Thesis Dissertation, Inst. Chem. Phys., Chernogolovka, 1978, 6 (in Russian).

158. **Maaraui, M. A., Nikipanchuk, M. V., Chernyak, B. I.,** Mechanism of chain generation at liquid-phase oxidation of crotonaldehyde in *n*-decane, *Kinet. Katal.,* 19, 499, 1978.

159. **Yanshilieva, N., Skibida, I. P., Maizus, Z.,** Mechanism of chain generation in oxidizing methyloleate and methyllinoliate, *Izv. Otd. Khim. Nauk Bolg. A.N.,* 4, 1, 1971.

160. **Dingchan, G. E., Khanukova, N. S., Vardanyan, R. L.,** The kinetics of oxidation of cholesterylformiate, *Arm. Khim. Zh.,* 30, 644, 1977.

161. **Sukhov, V. D., Mogilevich, M. M.,** Chain generation in oxidizing esters of acrylic and methacrylic acids, *Zh. Fiz. Khim.,* 53, 1477, 1979.

162. **Agisheva S. A.,** *The kinetics and mechanism of liquid-phase oxidation of 1,3-dioxacyclanes,* Cand. Sci. (Chem.) Thesis Dissertation,, Ufa, 1975, 18 (in Russian).

163. **Alexandrov, Yu. A., Sadikov, G. B., Chelnokova, I. L., Golov, V. G.,** On reaction of chain generation at autoxidation of 4,4'-diphenylmethanediisocyanate, *Zh. Org. Khim.,* 12, 855, 1976.

164. **Agisheva, S. A., Zlotskii, S. S., Imashev, U. B., Rakhmankulov, D. L.,** Generation of free radicals in oxidizing 1,3-dioxanes, *Izv. Vuzov, Ser. Khim. Khim. Technol.,* 21, 807, 1978.

Chapter 3

BOND DISSOCIATION ENERGIES AND RATE CONSTANTS OF REACTIONS OF PHENOLS

Table 3.1
O—H-Bond dissociation energies of phenols

No.	Name and formula of phenol	Mol. wt.	$D_{O—H}$/ kJ mol^{-1}	Ref.
1	Benzene, 1,3,5-tris(3′,5′-di-*tert*-butyl-4′-hydroxybenzyl);	742.20	344.9	1
	1,3,5-[3′,5′-$[(CH_3)_3C]_2$-4′-OH-$C_6H_5CH_2]_3C_6H_3$			
2	Benzoquinon, 2,6-bis-[3′,5′-di-*tert*-butyl-4′-hydroxyphenyl];	516.73	340.0	1
	2,6-[3′,5′-$[[(CH_3)_3C]_2C_6H_2$-4′-OH$]_2$-*cyclo*-[COCH=CHCOCH=CH]			
3	Catechol;	110.11	338.0	1
	2-HOC_6H_4OH			
4	Catechol, 4-*tert*-butyl;	150.22	342.4	1
	2-HO-4-$(CH_3)_3C$-C_6H_3OH			
5	Catechol, 3,5-di-*tert*-butyl;	206.33	338.3	1
	2-HO-3,5-$[(CH_3)_3C]_2C_6H_2OH$			
6	Catechol, 3,6-di-*tert*-butyl;	206.33	337.5	1
	2-HO-3,6-$[(CH_3)_3C]_2C_6H_2OH$			
7	*o*-Cresol;	108.14	357.5	1
	2-$CH_3C_6H_4OH$			
8	*m*-Cresol;	108.14	364.6	1
	3-$CH_3C_6H_4OH$			
9	*p*-Cresol;	108.14	357.8	1
	4-$CH_3C_6H_4OH$			
10	Chroman, 5,7,8-trimethyl-6-hydroxy;	192.26	328.8	1
	$(CH_3)_3C_6(OH)(CH_2)_3O$			

No.	Name and formula of phenol	Mol. wt.	D_{O-H}/ kJ mol^{-1}	Ref.
11	Chroman, 2,4,5,7-tetramethyl-6-hydroxy; $(CH_3)_2C_6H(OH)CH(CH_3)CH_2CH(CH_3)O$	206.29	324.4	1
12	Chroman, 2,5,7,8-tetramethyl-2-carboxyacetyl-6-hydroxy; $(CH_3)_3C_6(OH)(CH_2)_2CC(CH_3)(CH_2C(O)COOH)O$	306.36	331.1	1
13	Chroman, 2,5,7,8-tetramethyl-2-carboxy-6-hydroxy; $(CH_3)_3C_6(OH)(CH_2)_2C(CH_3)(COOH)O$	250.30	322.6	1
14	Chroman, 2,5,7,8-tetramethyl-2-carboxymethyl-6-hydroxy; $(CH_3)_3C_6(OH)(CH_2)_2C(CH_3)(CH_2COOH)O$	264.32	331.3	1
15	Chroman, 2,5,7,8-tetramethyl-2-(4′,8′-dimethylnonyl)-6-hydroxy; $(CH_3)_3C_6(OH)(CH_2)_2C(CH_3)(C_{11}H_{23})O$	344.58	327.5	1
16	Chroman, 2,5,7,8-tetramethyl-2-dimethylpropyl-6-hydroxy; $(CH_3)_3C_6(OH)(CH_2)_2C(CH_3)(C(CH_3)_2CH_2CH_3)O$	276.42	327.7	1
17	Chroman, 2,2,5,7,8-pentamethyl-6-hydroxy; $(CH_3)_3C_6(OH)(CH_2)_2C(CH_3)_2O$	220.31	326.4	1
18	Chroman, 2,3,5,7,8-pentamethyl-6-hydroxy; $(CH_3)_3C_6(OH)CH_2CH(CH_3)CH(CH_3)O$	220.31	327.3	1
19	Dihydrobenzofuran, 2,2,6,7-tetramethyl-5-hydroxy; $(CH_3)_2C_6H(OH)CH_2C(CH_3)_2O$	192.26	327.7	1
20	Dihydrobenzofuran, 2,4,6,7-tetramethyl-5-hydroxy; $(CH_3)_3C_6(OH)CH_2CH(CH_3)O$	189.23	324.4	1
21	Dihydrobenzofuran, 2,4,6,7-tetramethyl-2-carboxy-5-hydroxy; $(CH_3)_3C_6(OH)CH_2C(CH_3)(COOH)O$	233.25	332.0	1
22	Dihydrobenzofuran, 2,2,4,6,7-pentamethyl-5-hydroxy; $(CH_3)_3C_6(OH)CH_2C(CH_3)_2O$	206.29	324.0	1
23	3,4-Dihydrobenzopyrane, 4,4,5,7,8-penta-methyl-6-hydroxy; $(CH_3)_3C_6(OH)C(CH_3)_2(CH_2)_2O$	340.46	330.0	1
24	1,2-Ethane-2,2′-bis-(4,6-di-*tert*-butylphenol); 2,2′-$(CH_2)_2$- $[4,6\text{-}[(CH_3)_3C]_2C_6H_2OH]_2$	438.36	340.9	1
25	Fluorene, 1-hydroxy; 1-$HOC_{13}H_9$	182.22	336.3	1

No.	Name and formula of phenol	Mol. wt.	D_{O-H}/ kJ mol^{-1}	Ref.
26	Hydroquinone;	110.11	350.3	1
	4-HOC_6H_4OH			
27	4,4′-Methane-bis-(2,6-di-*tert*-butylphenol);	424.67	340.2	1
	4,4′-CH_2-[2,6-[$(CH_3)_3C]_2C_6H_2OH]_2$			
28	Naphthalene,1,5-dihydroxy;	164.20	333.5	1
	5-$HOC_{10}H_6OH$			
29	1-Naphthol;	144.17	341.4	1
	1-$HOC_{10}H_7$			
30	2-Naphthol;	144.17	351.8	1
	2-$HOC_{10}H_7$			
31	Pentaerithritol ester of 3,5-di-*tert*-butyl-4-hydroxyphenylpropionic acid;	1177.7	341.0	1
	$C[4-(CH_2)_2COOCH_2-2,6-[(CH_3)_3C]_2C_6H_2OH]_4$			
32	Phenantrene, 1-hydroxy;	139.22	352.7	1
	1-$HOC_{14}H_9$			
33	Phenantrene, 2-hydroxy;	139.22	365.0	1
	2-$HOC_{14}H_9$			
34	Phenantrene, 3-hydroxy;	139.22	360.5	1
	3-$HOC_{14}H_9$			
35	Phenantrene, 4-hydroxy;	139.22	354.2	1
	4-$HOC_{14}H_9$			
36	Phenol;	94.11	367.0	1
	C_6H_5OH			
37	Phenol, 3-acetyl;	136.15	368.1	2
	3-$CH_3COC_6H_4OH$			
38	Phenol, 4-acetyl;	136.15	369.5	2
	4-$CH_3COC_6H_4OH$			
39	Phenol, 3-amino;	109.13	365.2	2
	3-$NH_2C_6H_4OH$			
40	Phenol, 4-amino;	109.13	354.7	2
	4-$NH_2C_6H_4OH$			

No.	Name and formula of phenol	Mol. wt.	D_{O-H}/ kJ mol^{-1}	Ref.
41	Phenol, 4-benzyloxy; 4-$C_6H_5CH_2OC_6H_4OH$	200.24	347.5	1
42	Phenol, 2-bromo; 2-BrC_6H_4OH	173.01	369.8	2
43	Phenol, 4-bromo; 4-BrC_6H_4OH	173.01	370.6	3
44	Phenol, 4-butoxy; 4-$CH_3(CH_2)_3OC_6H_4OH$	166.22	345.8	1
45	Phenol, 2-*tert*-butyl; 2-$(CH_3)_3CC_6H_4OH$	150.22	351.8	1
46	Phenol, 4-*tert*-butyl; 4-$(CH_3)_3CC_6H_4OH$	150.22	356.9	1
47	Phenol, 2,4-di-*tert*-butyl; 2,4-$[(CH_3)_3C]_2C_6H_3OH$	206.33	357.5	1
48	Phenol, 2-*tert*-butyl-4-methyl; 2-$(CH_3)_3C$-4-$CH_3C_6H_3OH$	164.25	358.5	1
49	Phenol, 2,4-di-*tert*-butyl-6-methyl; 2,4-$[(CH_3)_3C]_2$-6-CH_3-C_6H_2OH	220.35	353.9	1
50	Diphenyl-[3,5,3′,5′-tetra-*tert*-butyl-4,4′-dihydroxy]; 4,4′-bis-[2,6-$[(CH_3)_3C]_2C_6H_2OH]_2$	410.64	342.6	1
51	Phenol, 2,6-di-*tert*-butyl; 2,6-$[(CH_3)_3C]_2C_6H_3OH$	206.33	346.7	1
52	Phenol, 2,4,6-tri-*tert*-butyl; 2,4,6-$[(CH_3)_3C]_3C_6H_2OH$	262.44	339.0	4
53	Phenol, 2,6-di-*tert*-butyl-4-acetyl; 2,6-$[(CH_3)_3C]_2$-4-$CH_3COC_6H_2OH$	248.37	345.5	1
54	Phenol, 2,6-di-*tert*-butyl-4-benzyl; 2,6-$[(CH_3)_3C]_2$-4-$C_6H_5CH_2C_6H_2OH$	296.45	339.7	1
55	Phenol, 2,6-di-*tert*-butyl-4-*tert*-butoxy; 2,6-$[(CH_3)_3C]_2$-4-$(CH_3)_3COC_6H_2OH$	262.44	331.3	1

No.	Name and formula of phenol	Mol. wt.	D_{O-H}/ kJ mol^{-1}	Ref.
56	Phenol, 2,6-di-*tert*-butyl-4-*tert*-butylcarboxylate; $2,6\text{-}[(CH_3)_3C]_2\text{-}4\text{-}(CH_3)_3COC(O)C_6H_2OH$	294.44	347.7	1
57	Phenol, 2,6-di-*tert*-butyl-4-carboxy; $2,6\text{-}[(CH_3)_3C]_2\text{-}4\text{-}HO(O)CC_6H_2OH$	250.34	348.8	1
58	Phenol, 2,6-di-*tert*-butyl-4-chloro; $2,6\text{-}[(CH_3)_3C]_2\text{-}4\text{-}ClC_6H_2OH$	240.77	344.5	1
59	Phenol, 2,6-di-*tert*-butyl-4-cumyl; $2,6\text{-}[(CH_3)_3C]_2\text{-}4\text{-}C_6H_5(CH_3)_2CC_6H_2OH$	324.51	340.1	1
60	Phenol, 2,6-di-*tert*-butyl-4-cyano; $2,6\text{-}[(CH_3)_3C]_2\text{-}4\text{-}NCC_6H_2OH$	231.34	352.4	1
61	Phenol, 2,6-di-*tert*-butyl-4-formyl; $2,6\text{-}[(CH_3)_3C]_2\text{-}4\text{-}H(O)CC_6H_2OH$	245.34	347.8	1
62	Phenol, 2,6-di-*tert*-butyl-4-methoxy; $2,6\text{-}[(CH_3)_3C]_2\text{-}4\text{-}CH_3OC_6H_2OH$	236.35	330.7	1
63	Phenol, 2,6-di-*tert*-butyl-4-methoxycarbonylmethyl; $2,6\text{-}[(CH_3)_3C]_2\text{-}4\text{-}CH_3OC(O)CH_2C_6H_2OH$	278.39	342.8	1
64	Phenol, 2,6-di-*tert*-butyl-4-methyl; $2,6\text{-}[(CH_3)_3C]_2\text{-}4\text{-}CH_3\text{-}C_6H_2OH$	220.35	339.0	1
65	Phenol, 2,6-di-*tert*-butyl-4-aminomethyl; $2,6\text{-}[(CH_3)_3C]_2\text{-}4\text{-}NH_2CH_2C_6H_2OH$	235.37	334.8	1
66	Phenol, 2,6-di-*tert*-butyl-4-carboxymethyl; $2,6\text{-}[(CH_3)_3C]_2\text{-}4\text{-}HO(O)CCH_2C_6H_2OH$	264.37	336.9	1
67	Phenol, 2,6-di-*tert*-butyl-4-(1′,1′,3′,3′-tetramethylbutyl); $2,6\text{-}[(CH_3)_3C]_2\text{-}4\text{-}(CH_3)_3CCH_2(CH_3)_2CC_6H_2OH$	318.54	339.8	1
68	Phenol, 2,6-di-*tert*-butyl-4-nitro; $2,6\text{-}[(CH_3)_3C]_2\text{-}4\text{-}O_2NC_6H_2OH$	251.33	358.0	1
69	Phenol, 2,6-di-*tert*-butyl-4-nitroso; $2,6\text{-}[(CH_3)_3C]_2\text{-}4\text{-}ONC_6H_2OH$	235.33	346.0	1
70	Phenol, 2,6-di-*tert*-butyl-4-phenyl; $2,6\text{-}[(CH_3)_3C]_2\text{-}4\text{-}C_6H_5C_6H_2OH$	282.43	337.7	1

No.	Name and formula of phenol	Mol. wt.	D_{O-H}/ kJ mol^{-1}	Ref.
71	Phenol, 2,6-di-*tert*-butyl-4-diphenylmethyl; $2,6\text{-}[(CH_3)_3C]_2\text{-}4\text{-}(C_6H_5)_2CHC_6H_2OH$	372.55	342.3	1
72	Phenol, 2,6-di-*tert*-butyl-4-octadecyloxycarbonylethyl; $2,6\text{-}[(CH_3)_3C]_2\text{-}4\text{-}C_{18}H_{37}OC(O)CH_2CH_2C_6H_2OH$	498.87	339.6	1
73	Phenol, 2,6-di-*tert*-butyl-4-thiophenyl; $2,6\text{-}[(CH_3)_3C]_2\text{-}4\text{-}C_6H_5SC_6H_2OH$	314.49	346.4	1
74	Phenol, 2-*tert*-butyl-4,6-dimethyl; $2\text{-}(CH_3)_3C\text{-}4,6\text{-}(CH_3)_2C_6H_2OH$	178.27	353.8	1
75	Phenol, 2-*tert*-butyl-4-methoxy; $2\text{-}(CH_3)_3C\text{-}4\text{-}CH_3OC_6H_3OH$	180.25	341.4	1
76	Phenol, 2-methoxy; $2\text{-}CH_3OC_6H_4OH$	124.14	357.8	1
77	Phenol, 3-methoxy; $3\text{-}CH_3OC_6H_4OH$	124.14	367.6	1
78	Phenol, 4-methoxy; $4\text{-}CH_3OC_6H_4OH$	124.14	349.0	1
79	Phenol, 2,3-dimethyl; $2,3\text{-}(CH_3)_2C_6H_3OH$	122.17	353.5	1
80	Phenol, 3,4-dimethyl; $3,4\text{-}(CH_3)_2C_6H_3OH$	122.17	352.6	1
81	Phenol, 3,5-dimethyl; $3,5\text{-}(CH_3)_2C_6H_3OH$	122.17	354.0	1
82	Phenol, 2,4-dimethyl; $2,4\text{-}(CH_3)_2C_6H_3OH$	122.17	358.5	1
83	Phenol, 2,6-dimethyl; $2,6\text{-}(CH_3)_2C_6H_3OH$	122.17	354.9	1
84	Phenol, 3-dimethylamino; $3\text{-}(CH_3)_2NC_6H_4OH$	123.20	358.6	3
85	Phenol, 4-dimethylamino; $4\text{-}(CH_3)_2NC_6H_4OH$	123.20	326.8	3

No.	Name and formula of phenol	Mol. wt.	D_{O-H}/ kJ mol^{-1}	Ref.
86	Phenol, 3-nitro; $3\text{-}NO_2C_6H_4OH$	139.11	365.1	2
87	Phenol, 4-nitro; $4\text{-}NO_2C_6H_4OH$	139.11	369.7	1
88	Phenol, 4-phenyl; $4\text{-}C_6H_5C_6H_4OH$	170.21	357.4	3
89	Phenol, 2,6-diphenyl-4-aminomethyl; $2,6\text{-}(C_6H_5)_2\text{-}4\text{-}NH_2CH_2C_6H_2OH$	276.36	322.9	1
90	Phenol, 2,6-diphenyl-4-cyanomethyl; $2,6\text{-}(C_6H_5)_2\text{-}4\text{-}CNCH_2C_6H_2OH$	273.34	311.1	
91	Phenol, 2,6-diphenyl-4-carboxymethyl; $2,6\text{-}(C_6H_5)_2\text{-}4\text{-}HO(O)CCH_2C_6H_2OH$	304.35	325.9	1
92	Phenol, 2,6-di-(1′-phenyl)ethyl-4-mercapto; $2,6\text{-}[CH(CH_3)C_6H_5]_2\text{-}4\text{-}HSC_6H_2OH$	334.48	340.0	1
93	Phenol, 2,6-di-(1′-phenyl)ethyl-4-sulfobutyl; $2,6\text{-}[CH(CH_3)C_6H_5]_2\text{-}4\text{-}CH_3(CH_2)_3SC_6H_2OH$	390.59	341.2	1
94	Phenol, 2,6-di-(1′-phenyl)ethyl-4-sulfomethyl; $2,6\text{-}[CH(CH_3)C_6H_5]_2\text{-}4\text{-}CH_3SC_6H_2OH$	348.51	340.4	1
95	Phenol, 2,6-diphenyl-4-methoxy; $2,6\text{-}(C_6H_5)\text{-}4\text{-}CH_3OC_6H_2OH$	276.34	328.0	1
96	Phenol, 2,6-diphenyl-4-octadecyloxy; $2,6\text{-}(C_6H_5)_2\text{-}4\text{-}C_{18}H_{37}OC_6H_2OH$	514.80	328.6	1
97	Phenol, 2,6-diphenyl-4-tetracozyloxy; $2,6\text{-}(C_6H_5)_2\text{-}4\text{-}C_{24}H_{49}OC_6H_2OH$	598.96	329.2	1
98	Phenol, 2-sulfobenzyl-4-methyl-6-(1′-phenyl)ethyl; $2\text{-}SCH_2C_6H_5\text{-}4\text{-}CH_3\text{-}6\text{-}CH_3(C_6H_5)CHC_6H_2OH$	334.48	338.0	1
99	Phenol , 2-sulfobutyl-4-methyl-6-(1′-phenyl)ethyl; $2\text{-}SC_4H_9\text{-}4\text{-}CH_3\text{-}6\text{-}CH_3(C_6H_5)CHC_6H_2OH$	300.46	338.5	1
100	Phenol, 2-sulfoethyl,4-methyl-6-(1′-phenyl)ethyl; $2\text{-}SCH_2CH_3\text{-}4\text{-}CH_3\text{-}6\text{-}CH_3(C_6H_5)CHC_6H_2OH$	272.41	338.7	1

No.	Name and formula of phenol	Mol. wt.	D_{O-H}/ kJ mol^{-1}	Ref.
101	Phenol, 4-carboxy; $4\text{-}HO(O)CC_6H_4OH$	138.12	369.7	1
102	Phenol, 4-chloro; $4\text{-}ClC_6H_4OH$	128.56	368.6	1
103	Phenol, 3-chloro; $3\text{-}ClC_6H_4OH$	128.56	375.4	3
104	Phenol, 2-chloro; $2\text{-}ClC_6H_4OH$	128.56	367.5	3
105	Phenol, 3-cyano; $3\text{-}CNC_6H_4OH$	119.12	371.6	2
106	Phenol, 4-cyano; $4\text{-}CNC_6H_4OH$	119.12	368.2	2
107	Phenol, 2-(2′-cyano)sulfoethyl-4-methyl-6-(1′-phenyl)ethyl; $2\text{-}S(CH_2)_2CN\text{-}4\text{-}CH_3\text{-}6\text{-}CH_3(C_6H_5)CHC_6H_2OH$	297.42	345.5	1
108	Phenol, 2-methyl,4-*tert*-butyl; $2\text{-}CH_3\text{-}4\text{-}(CH_3)_3C\text{-}C_6H_3OH$	164.25	357.9	1
109	Phenol, 2,4,5-tri-methyl; $2,4,5\text{-}(CH_3)_3C_6H_2OH$	136.19	354.8	1
110	Phenol, 2,4,6-tri-methyl; $2,4,6\text{-}(CH_3)_3C_6H_2OH$	136.19	346.8	1
111	Phenol, 2-methyl-4-aminomethyl-6-*tert*-butyl; $2\text{-}CH_3\text{-}4\text{-}NH_2CH_2\text{-}6\text{-}(CH_3)_3CC_6H_2OH$	193.29	348.6	1
112	Phenol, 2,6-dimethyl-4-aminomethyl; $2,6\text{-}(CH_3)_2\text{-}4\text{-}NH_2CH_2\text{-}C_6H_2OH$	151.21	346.1	1
113	Phenol, 2,6-dimethyl-4-carboxymethyl; $2,6\text{-}(CH_3)_2\text{-}4\text{-}HOC(O)CH_2C_6H_2OH$	168.19	348.0	1
114	Phenol, 2,6-dimethyl-4-cyanomethyl; $2,6\text{-}(CH_3)_2\text{-}4\text{-}CNCH_2\text{-}C_6H_2OH$	149.19	352.0	1
115	Phenol, 2,3,4,6-tetramethyl; $2,3,4,6\text{-}(CH_3)_4\text{-}C_6HOH$	150.22	345.4	1

No.	Name and formula of phenol	Mol. wt.	D_{O-H}/ kJ mol^{-1}	Ref.
116	Phenol, 2,3,5,6-tetramethyl;	150.22	349.2	1
	2,3,5,6-$(CH_3)_4C_6HOH$			
117	Phenol, pentamethyl;	164.25	338.5	1
	$(CH_3)_5C_6OH$			
118	Phenol, 2,3,5,6-tetramethyl-4-methoxy;	180.25	338.1	1
	2,3,5,6-$(CH_3)_4$-4-CH_3OC_6OH			
119	Pentaeritrityl tetrakis[3-(3,5-di-*tert*-butyl-4-hydroxyphenyl)propionate;	1177.70	339.7	1
	[3,5-$[(CH_3)_3C]_2$-4-$[HOC_6H_2(CH_2)_2COOCH_2]_4C$			
120	Ethyleneglycol bis(3,5-di-*tert*-butyl-4-hydroxyphenyl)propionate;	622.84	338.9	1
	[2,6-$[(CH_3)_3C]_2$-4-$(CH_2)_2COO(CH_2)_2C_6H_2OH]_2O$			
121	Resorcinol;	110.11	365.8	1
	3-HOC_6H_4OH			
122	Silane, tetra(ethyloxy-3,5-di-*tert*-butyl-4-hydroxyphenyl);	897.56	340.9	1
	Si[2,6-$[(CH_3)_3C]_2$-4-$CH_2CH_2OC_6H_2OH]_4$			
123	4,4′-Sulfide-bis-(3,5-di-*tert*-butyl-4-hydroxybenzyl);	470.76	339.9	1
	4,4′-CH_2SCH_2-[2,6-$[(CH_3)_3C]_2C_6H_2OH]_2$			
124	2,2′-Sulfide-bis-4-methyl-5-hydroxy-6-(1′-phenyl)ethyl;	274.47	335.2	1
	2,2′-S-[4-CH_3-6-$CH_3(C_6H_5)CHC_6H_2OH]_2$			
125	2,2′-Disulfide-bis-[4-methyl-6-hydroxy-5-(1′-phenyl)ethyl;	486.70	334.6	1
	2,2′-SS-[4-CH_3-6-$CH(CH_3)(C_6H_5)C_6H_2OH]_2$			
126	Indophenol;	410.54	327.0	5
	cyclo-$[C(O)C(C(CH_3)_3)CHC{=}N(-)CHC(C(CH_3)_3]$-2,6-$(C(CH_3)_3)_2$-4-$C_6H_2OH$			
127	Tetrahydroquinoline, 5,7,8-trimethyl-6-hydroxy-N-acetamido;	217.31	333.7	1
	$(CH_3)_3C_6(OH)(CH_2)_3NC(O)CH_3$			
128	Tetrahydroquinoline, 5,7,8-trimethyl-6-hydroxy-N-ethyl;	219.33	340.3	1
	$(CH_3)_3C_6(OH)(CH_2)_3NCH_2CH_3$			
129	Tocol, 5,7-dimethyl;	391.70	331.3	1
	$(CH_3)_2C_6H(OH)(CH_2)_2OC(CH_3)CH_2[(CH_2)_2CH(CH_3)CH_2]_3H$			
130	α-Tocopherol;	430.72	327.7	1
	α-$(CH_3)_3C_6(OH)(CH_2)_2OC(CH_3)[(CH_2)_3CH(CH_3)]_3CH_3$			

No.	Name and formula of phenol	Mol. wt.	D_{O-H}/ kJ mol^{-1}	Ref.
131	β-Tocopherol;	416.69	332.4	1
	β-$(CH_3)_2C_6H(OH)(CH_2)_2OC(CH_3)[(CH_2)_3CH(CH_3)]_3CH_3$			
132	γ-Tocopherol;	416.69	321.7	1
	γ-$(CH_3)_2C_6H(OH)(CH_2)_2OC(CH_3)[(CH_2)_3CH(CH_3)]_3CH_3$			
133	δ-Tocopherol;	402.66	317.0	1
	δ-$CH_3C_6H_2(OH)(CH_2)_2OC(CH_3)[(CH_2)_3CH(CH_3)]_3CH_3$			

Table 3.2
Enthalpies, activation energies and rate constants of reactions of peroxyl radicals with phenols (Ar_1OH), in hydrocarbon solutions calculated by formulas 1.15–1.17 and 1.21. The values of A, br_e and α, see Table 1.6

No.	Phenol	$RO_2^•$	ΔH/ kJ mol^{-1}	E/ kJ mol^{-1}	k (333 K)/ l mol^{-1} s^{-1}	k (400 K)/ l mol^{-1} s^{-1}
3	2-HOC_6H_4OH	$HO_2^•$	−31.0	11.0	6.0×10^5	1.2×10^6
3	2-HOC_6H_4OH	*sec*-$RO_2^•$	−27.5	12.5	3.5×10^5	7.5×10^5
3	2-HOC_6H_4OH	*tert*-$RO_2^•$	−20.6	15.5	1.2×10^5	3.0×10^5
4	2-HO-4-$(CH_3)_3CC_6H_3OH$	$HO_2^•$	−26.6	12.9	3.0×10^5	6.6×10^5
4	2-HO-4-$(CH_3)_3CC_6H_3OH$	*sec*-$RO_2^•$	−23.1	14.4	1.8×10^5	4.2×10^5
4	2-HO-4-$(CH_3)_3CC_6H_3OH$	*tert*-$RO_2^•$	−16.2	17.5	5.8×10^4	1.7×10^5
9	4-$CH_3C_6H_4OH$	$HO_2^•$	−11.2	19.8	2.5×10^4	8.3×10^4
9	4-$CH_3C_6H_4OH$	*sec*-$RO_2^•$	−7.7	21.4	1.4×10^4	5.1×10^4
9	4-$CH_3C_6H_4OH$	*tert*-$RO_2^•$	−0.8	24.8	4.1×10^3	1.8×10^4
10	$(CH_3)_3C_6(OH)(CH_2)_3O$	$HO_2^•$	−40.2	7.3	2.3×10^6	3.6×10^6
10	$(CH_3)_3C_6(OH)(CH_2)_3O$	*sec*-$RO_2^•$	−36.7	8.7	1.4×10^6	2.3×10^6
10	$(CH_3)_3C_6(OH)(CH_2)_3O$	*tert*-$RO_2^•$	−29.8	11.5	5.0×10^5	1.0×10^6
20	$(CH_3)_3C_6(OH)CH_2CH(CH_3)O$	$HO_2^•$	−44.6	5.6	4.2×10^6	5.9×10^6
20	$(CH_3)_3C_6(OH)CH_2CH(CH_3)O$	*sec*-$RO_2^•$	−41.1	7.0	2.6×10^6	3.9×10^6
20	$(CH_3)_3C_6(OH)CH_2CH(CH_3)O$	*tert*-$RO_2^•$	−34.2	9.7	9.6×10^5	1.7×10^6
25	1-$HOC_{13}H_9$	$HO_2^•$	−32.7	10.3	7.8×10^5	1.4×10^6

No.	Phenol	$RO_2^{\bullet}$	ΔH/ kJ mol^{-1}	E/ kJ mol^{-1}	k (333 K)/ l mol^{-1} s^{-1}	k (400 K)/ l mol^{-1} s^{-1}
25	1-$HOC_{13}H_9$	*sec*-$RO_2^{\bullet}$	−29.2	11.8	4.5×10^5	9.2×10^5
25	1-$HOC_{13}H_9$	*tert*-$RO_2^{\bullet}$	−22.3	14.7	1.6×10^5	3.8×10^5
29	1-$HOC_{10}H_7$	$HO_2^{\bullet}$	−27.6	12.4	3.6×10^5	7.7×10^5
29	1-$HOC_{10}H_7$	*sec*-$RO_2^{\bullet}$	−24.1	13.9	2.1×10^5	4.9×10^5
29	1-$HOC_{10}H_7$	*tert*-$RO_2^{\bullet}$	−17.2	17.0	7.0×10^4	1.9×10^5
30	2-$HOC_{10}H_7$	$HO_2^{\bullet}$	−17.2	17.0	7.0×10^4	1.9×10^5
30	2-$HOC_{10}H_7$	*sec*-$RO_2^{\bullet}$	−13.7	18.6	3.9×10^4	1.2×10^5
30	2-$HOC_{10}H_7$	*tert*-$RO_2^{\bullet}$	−6.8	21.9	1.2×10^4	4.4×10^4
32	1-$HOC_{14}H_9$	$HO_2^{\bullet}$	−16.3	17.4	6.0×10^4	1.7×10^5
32	1-$HOC_{14}H_9$	*sec*-$RO_2^{\bullet}$	−12.8	19.0	3.3×10^4	1.1×10^5
32	1-$HOC_{14}H_9$	*tert*-$RO_2^{\bullet}$	−5.9	22.3	1.0×10^4	3.9×10^4
36	C_6H_5OH	$HO_2^{\bullet}$	−2.0	24.2	5.1×10^3	2.2×10^4
36	C_6H_5OH	*sec*-$RO_2^{\bullet}$	1.5	25.9	2.8×10^3	1.3×10^4
36	C_6H_5OH	*tert*-$RO_2^{\bullet}$	8.4	29.5	7.5×10^2	4.5×10^3
38	4-$CH_3COC_6H_4OH$	$HO_2^{\bullet}$	0.5	25.4	3.3×10^3	1.5×10^4
38	4-$CH_3COC_6H_4OH$	*sec*-$RO_2^{\bullet}$	4.0	27.2	1.7×10^3	9.0×10^3
38	4-$CH_3COC_6H_4OH$	*tert*-$RO_2^{\bullet}$	10.9	30.8	4.7×10^2	3.0×10^3
40	4-$NH_2C_6H_4OH$	$HO_2^{\bullet}$	−14.3	18.3	4.3×10^4	1.3×10^5
40	4-$NH_2C_6H_4OH$	*sec*-$RO_2^{\bullet}$	−10.8	20.0	2.3×10^4	7.8×10^4
40	4-$NH_2C_6H_4OH$	*tert*-$RO_2^{\bullet}$	−3.9	23.3	7.1×10^3	2.9×10^4
41	4-$C_6H_5CH_2OC_6H_4OH$	$HO_2^{\bullet}$	−21.5	15.1	1.4×10^5	3.4×10^5
41	4-$C_6H_5CH_2OC_6H_4OH$	*sec*-$RO_2^{\bullet}$	−18.0	16.6	8.0×10^4	2.2×10^5
41	4-$C_6H_5CH_2OC_6H_4OH$	*tert*-$RO_2^{\bullet}$	−11.1	19.8	2.5×10^4	8.3×10^4
44	4-$CH_3(CH_2)_3OC_6H_4OH$	$HO_2^{\bullet}$	−23.2	14.3	1.8×10^5	4.3×10^5
44	4-$CH_3(CH_2)_3OC_6H_4OH$	*sec*-$RO_2^{\bullet}$	−19.7	15.9	1.0×10^5	2.7×10^5
44	4-$CH_3(CH_2)_3OC_6H_4OH$	*tert*-$RO_2^{\bullet}$	−12.8	19.0	3.3×10^4	1.1×10^5
46	4-$(CH_3)_3CC_6H_4OH$	$HO_2^{\bullet}$	−12.1	19.3	3.0×10^4	9.7×10^4
46	4-$(CH_3)_3CC_6H_4OH$	*sec*-$RO_2^{\bullet}$	−8.6	21.0	1.6×10^4	5.8×10^4

No.	Phenol	$RO_2^\bullet$	ΔH/ kJ mol^{-1}	E/ kJ mol^{-1}	k (333 K)/ l mol^{-1} s^{-1}	k (400 K)/ l mol^{-1} s^{-1}
46	$4\text{-}(CH_3)_3CC_6H_4OH$	*tert*-$RO_2^\bullet$	−1.7	24.3	4.9×10^3	2.1×10^4
47	$2,4\text{-}[(CH_3)_3C]_2C_6H_3OH$	$HO_2^\bullet$	−11.5	19.6	2.7×10^4	8.8×10^4
47	$2,4\text{-}[(CH_3)_3C]_2C_6H_3OH$	*sec*-$RO_2^\bullet$	−8.0	21.3	1.5×10^4	5.3×10^4
47	$2,4\text{-}[(CH_3)_3C]_2C_6H_3OH$	*tert*-$RO_2^\bullet$	−1.1	24.6	4.4×10^3	2.0×10^4
74	$2\text{-}(CH_3)_3C\text{-}4,6\text{-}(CH_3)_2C_6H_2OH$	$HO_2^\bullet$	−15.2	17.9	5.0×10^4	1.5×10^5
74	$2\text{-}(CH_3)_3C\text{-}4,6\text{-}(CH_3)_2C_6H_2OH$	*sec*-$RO_2^\bullet$	−11.7	19.5	2.8×10^4	9.1×10^4
74	$2\text{-}(CH_3)_3C\text{-}4,6\text{-}(CH_3)_2C_6H_2OH$	*tert*-$RO_2^\bullet$	−4.8	22.8	8.5×10^3	3.4×10^4
75	$2\text{-}(CH_3)_3C\text{-}4\text{-}CH_3OC_6H_3OH$	$HO_2^\bullet$	−27.6	12.4	3.6×10^5	7.7×10^5
75	$2\text{-}(CH_3)_3C\text{-}4\text{-}CH_3OC_6H_3OH$	*sec*-$RO_2^\bullet$	−24.1	13.9	2.1×10^5	4.9×10^5
75	$2\text{-}(CH_3)_3C\text{-}4\text{-}CH_3OC_6H_3OH$	*tert*-$RO_2^\bullet$	−17.2	17.0	6.9×10^4	1.9×10^5
78	$4\text{-}CH_3OC_6H_4OH$	$HO_2^\bullet$	−20.0	15.7	1.1×10^5	2.9×10^5
78	$4\text{-}CH_3OC_6H_4OH$	*sec*-$RO_2^\bullet$	−16.5	17.3	6.2×10^4	1.8×10^5
78	$4\text{-}CH_3OC_6H_4OH$	*tert*-$RO_2^\bullet$	−9.6	20.5	1.9×10^4	6.7×10^4
79	$2,3\text{-}(CH_3)_2C_6H_3OH$	$HO_2^\bullet$	−15.5	17.8	5.2×10^4	1.5×10^5
79	$2,3\text{-}(CH_3)_2C_6H_3OH$	*sec*-$RO_2^\bullet$	−12.0	19.4	2.9×10^4	9.4×10^4
79	$2,3\text{-}(CH_3)_2C_6H_3OH$	*tert*-$RO_2^\bullet$	−5.1	22.7	8.8×10^3	3.5×10^4
80	$3,4\text{-}(CH_3)_2C_6H_3OH$	$HO_2^\bullet$	−16.4	17.4	6.0×10^4	1.7×10^5
80	$3,4\text{-}(CH_3)_2C_6H_3OH$	*sec*-$RO_2^\bullet$	−12.9	19.0	3.3×10^4	1.1×10^5
80	$3,4\text{-}(CH_3)_2C_6H_3OH$	*tert*-$RO_2^\bullet$	−6.0	22.2	1.1×10^4	4.0×10^4
81	$3,5\text{-}(CH_3)_2C_6H_3OH$	$HO_2^\bullet$	−15.0	18.0	4.8×10^4	1.4×10^5
81	$3,5\text{-}(CH_3)_2C_6H_3OH$	*sec*-$RO_2^\bullet$	−11.5	19.6	2.7×10^4	8.8×10^4
81	$3,5\text{-}(CH_3)_2C_6H_3OH$	*tert*-$RO_2^\bullet$	−4.6	22.9	8.2×10^3	3.3×10^4
82	$2,4\text{-}(CH_3)_2C_6H_3OH$	$HO_2^\bullet$	−10.5	20.0	2.3×10^4	7.8×10^4
82	$2,4\text{-}(CH_3)_2C_6H_3OH$	*sec*-$RO_2^\bullet$	−7.0	21.8	1.2×10^4	4.6×10^4
82	$2,4\text{-}(CH_3)_2C_6H_3OH$	*tert*-$RO_2^\bullet$	−0.1	25.2	3.6×10^3	1.6×10^4
83	$2,6\text{-}(CH_3)_2C_6H_3OH$	$HO_2^\bullet$	−14.1	18.4	4.2×10^4	1.3×10^5
83	$2,6\text{-}(CH_3)_2C_6H_3OH$	*sec*-$RO_2^\bullet$	−10.6	20.0	2.3×10^4	7.8×10^4
83	$2,6\text{-}(CH_3)_2C_6H_3OH$	*tert*-$RO_2^\bullet$	−3.7	23.3	7.1×10^3	2.9×10^4

No.	Phenol	$RO_2^{\bullet}$	ΔH/ kJ mol^{-1}	E/ kJ mol^{-1}	k (333 K)/ l mol^{-1} s^{-1}	k (400 K)/ l mol^{-1} s^{-1}
85	4-$(CH_3)_2NC_6H_4OH$	$HO_2^{\bullet}$	−42.2	6.6	3.0×10^6	4.4×10^6
85	4-$(CH_3)_2NC_6H_4OH$	*sec*-$RO_2^{\bullet}$	−38.7	7.9	1.8×10^6	3.0×10^6
85	4-$(CH_3)_2NC_6H_4OH$	*tert*-$RO_2^{\bullet}$	−31.8	10.7	6.7×10^5	1.3×10^6
110	2,4,6-$(CH_3)_3C_6H_2OH$	$HO_2^{\bullet}$	−22.2	14.8	1.5×10^5	3.8×10^5
110	2,4,6-$(CH_3)_3C_6H_2OH$	*sec*-$RO_2^{\bullet}$	−18.7	16.3	8.9×10^4	2.4×10^5
110	2,4,6-$(CH_3)_3C_6H_2OH$	*tert*-$RO_2^{\bullet}$	−11.8	19.5	2.8×10^4	9.1×10^4
130	α-Tocopherol	$HO_2^{\bullet}$	−41.3	6.9	2.6×10^6	4.0×10^6
130	α-Tocopherol	*sec*-$RO_2^{\bullet}$	−37.8	8.2	1.7×10^6	2.7×10^6
130	α-Tocopherol	*tert*-$RO_2^{\bullet}$	−30.9	11.0	6.0×10^5	1.2×10^6
131	β-Tocopherol	$HO_2^{\bullet}$	−36.6	8.7	1.4×10^6	2.3×10^6
131	β-Tocopherol	*sec*-$RO_2^{\bullet}$	−33.1	10.2	8.0×10^5	1.5×10^6
131	β-Tocopherol	*tert*-$RO_2^{\bullet}$	−26.2	13.0	2.9×10^5	6.4×10^5
133	δ-Tocopherol	$HO_2^{\bullet}$	−52.0	2.9	1.1×10^7	1.3×10^7
133	δ-Tocopherol	*sec*-$RO_2^{\bullet}$	−48.5	4.2	7.0×10^6	9.0×10^6
133	δ-Tocopherol	*tert*-$RO_2^{\bullet}$	−41.6	6.8	2.7×10^6	4.1×10^6

Table 3.3
Enthalpies, activation energies and rate constants of reaction of peroxyl radicals with sterically hindered phenols (Ar_2OH) in hydrocarbon solutions calculated by formulas 1.15–1.17 and 1.21. The values of *A*, br_e and α, see Table 1.6

No.	Phenol	$RO_2^{\bullet}$	ΔH/ kJ mol^{-1}	E/ kJ mol^{-1}	k(333 K)/ l mol^{-1} s^{-1}	k(400 K)/ l mol^{-1} s^{-1}
1	1,3,5-[3′,5′-$[(CH_3)_3C]_2$-4′-HO-$C_6H_2CH_2]_3C_6H_3$	$HO_2^{\bullet}$	−24.1	20.4	6.0×10^4	2.1×10^5
1	1,3,5-[3′,5′-$[(CH_3)_3C]_2$-4′-HO-$C_6H_2CH_2]_3C_6H_3$	*sec*-$RO_2^{\bullet}$	−20.6	22.0	3.3×10^4	1.3×10^5
1	1,3,5-[3′,5′-$[(CH_3)_3C]_2$-4′-HO-$C_6H_3CH_2]_3C_6H_3$	*tert*-$RO_2^{\bullet}$	−13.7	25.1	1.1×10^4	5.1×10^4
24	2,2′-CH_2-[4,6-$[(CH_3)_3C]_2C_6H_2OH]_2$	$HO_2^{\bullet}$	−28.1	18.6	7.4×10^4	2.4×10^5
24	2,2′-CH_2-[4,6-$[(CH_3)_3C]_2C_6H_2OH]_2$	*sec*-$RO_2^{\bullet}$	−24.6	20.2	4.4×10^4	1.5×10^5
24	2,2′-CH_2-[4,6-$[(CH_3)_3C]_2C_6H_2OH]_2$	*tert*-$RO_2^{\bullet}$	−17.7	23.3	1.4×10^4	5.8×10^4
31	C-[4-$(CH_2)_2COOCH_2$-2,6-$[(CH_3)_3C]_2C_6H_2OH]$	$HO_2^{\bullet}$	−28.0	18.7	3.7×10^4	1.2×10^5

No.	Phenol	$RO_2^•$	ΔH/ kJ mol^{-1}	E/ kJ mol^{-1}	k(333 K)/ l mol^{-1} s^{-1}	k(400 K)/ l mol^{-1} s^{-1}
31	C-[4-$(CH_2)_2COOCH_2$-2,6-$[(CH_3)_3C]_2C_6H_2OH$]	*sec*-$RO_2^•$	−24.5	20.2	2.2×10^4	7.4×10^4
31	C-[4-$(CH_2)_2COOCH_2$-2,6-$[(CH_3)_3C]_2C_6H_2OH$]	*tert*-$RO_2^•$	−17.6	23.3	7.1×10^3	2.9×10^4
49	2,4-$[(CH_3)_3C]_2$-6-CH_3-C_6H_2OH	$HO_2^•$	−15.1	24.5	4.6×10^3	2.0×10^4
49	2,4-$[(CH_3)_3C]_2$-6-CH_3-C_6H_2OH	*sec*-$RO_2^•$	−11.6	26.1	2.6×10^3	1.2×10^4
49	2,4-$[(CH_3)_3C]_2$-6-CH_3-C_6H_2OH	*tert*-$RO_2^•$	−4.7	29.4	7.8×10^2	4.6×10^3
50	4,4′-[2,6-$[(CH_3)_3C]_2C_6H_2OH]_2$	$HO_2^•$	−26.4	19.4	5.8×10^4	1.9×10^5
50	4,4′-[2,6-$[(CH_3)_3C]_2C_6H_2OH]_2$	*sec*-$RO_2^•$	−22.9	20.9	3.4×10^4	1.2×10^5
50	4,4′-[2,6-$[(CH_3)_3C]_2C_6H_2OH]_2$	*tert*-$RO_2^•$	−16.0	24.0	1.1×10^4	4.8×10^4
51	2,6-$[(CH_3)_3C]_2C_6H_3OH$	$HO_2^•$	−22.3	21.2	1.5×10^4	5.5×10^4
51	2,6-$[(CH_3)_3C]_2C_6H_3OH$	*sec*-$RO_2^•$	−18.8	22.8	8.5×10^3	3.4×10^4
51	2,6-$[(CH_3)_3C]_2C_6H_3OH$	*tert*-$RO_2^•$	−11.9	26.0	2.7×10^3	1.3×10^4
52	2,6-$[(CH_3)_3C]_2$-4-$CH_3CONHC_6H_2OH$	$HO_2^•$	−42.3	12.7	3.3×10^5	7.0×10^5
52	2,6-$[(CH_3)_3C]_2$-4-$CH_3CONHC_6H_2OH$	*sec*-$RO_2^•$	−38.8	14.2	1.9×10^5	4.5×10^5
52	2,6-$[(CH_3)_3C]_2$-4-$CH_3CONHC_6H_2OH$	*tert*-$RO_2^•$	−31.9	17.0	6.9×10^4	1.9×10^5
53	2,6-$[(CH_3)_3C]_2$-4-$CH_3COC_6H_2OH$	$HO_2^•$	−23.5	20.7	1.8×10^4	6.3×10^4
53	2,6-$[(CH_3)_3C]_2$-4-$CH_3COC_6H_2OH$	*sec*-$RO_2^•$	−20.0	22.2	1.1×10^4	4.0×10^4
53	2,6-$[(CH_3)_3C]_2$-4-$CH_3COC_6H_2OH$	*tert*-$RO_2^•$	−13.1	25.4	3.3×10^3	1.5×10^4
54	2,6-$[(CH_3)_3C]_2$-4-$C_6H_5CH_2C_6H_2OH$	$HO_2^•$	−29.3	18.1	4.6×10^4	1.4×10^5
54	2,6-$[(CH_3)_3C]_2$-4-$C_6H_5CH_2C_6H_2OH$	*sec*-$RO_2^•$	−25.8	19.6	2.7×10^4	8.8×10^4
54	2,6-$[(CH_3)_3C]_2$-4-$C_6H_5CH_2C_6H_2OH$	*tert*-$RO_2^•$	−18.9	22.7	8.8×10^3	3.5×10^4
55	2,6-$[(CH_3)_3C]_2$-4-$(CH_3)_3COC_6H_2OH$	$HO_2^•$	−37.7	14.6	1.6×10^5	4.0×10^5
55	2,6-$[(CH_3)_3C]_2$-4-$(CH_3)_3COC_6H_2OH$	*sec*-$RO_2^•$	−34.2	16.1	9.5×10^4	2.5×10^5
55	2,6-$[(CH_3)_3C]_2$-4-$(CH_3)_3COC_6H_2OH$	*tert*-$RO_2^•$	−27.3	19.0	3.3×10^4	1.1×10^5
56	2,6-$[(CH_3)_3C]_2$-4-$(CH_3)_3COC(O)C_6H_2OH$	$HO_2^•$	−21.3	21.6	1.3×10^4	4.8×10^4
56	2,6-$[(CH_3)_3C]_2$-4-$(CH_3)_3COC(O)C_6H_2OH$	*sec*-$RO_2^•$	−17.8	23.2	7.3×10^3	3.0×10^4
56	2,6-$[(CH_3)_3C]_2$-4-$(CH_3)_3COC(O)C_6H_2OH$	*tert*-$RO_2^•$	−10.9	26.4	2.3×10^3	1.1×10^4
57	2,6-$[(CH_3)_3C]_2$-4-$HOOCC_6H_2OH$	$HO_2^•$	−20.2	22.1	1.1×10^4	4.2×10^4
57	2,6-$[(CH_3)_3C]_2$-4-$HOOCC_6H_2OH$	*sec*-$RO_2^•$	−16.7	23.7	6.1×10^3	2.6×10^4

No.	Phenol	$RO_2^•$	ΔH/ kJ mol^{-1}	E/ kJ mol^{-1}	k(333 K)/ l mol^{-1} s^{-1}	k(400 K)/ l mol^{-1} s^{-1}
57	2,6-$[(CH_3)_3C]_2$-4-$HOOCC_6H_2OH$	*tert*-$RO_2^•$	− 9.8	27.0	1.9×10^3	9.5×10^3
58	2,6-$[(CH_3)_3C]_2$-4-ClC_6H_2OH	$HO_2^•$	−24.5	20.2	2.2×10^4	7.4×10^4
58	2,6-$[(CH_3)_3C]_2$-4-ClC_6H_2OH	*sec*-$RO_2^•$	−21.0	21.8	1.2×10^4	4.6×10^4
58	2,6-$[(CH_3)_3C]_2$-4-ClC_6H_2OH	*tert*-$RO_2^•$	−14.1	24.9	4.0×10^3	1.8×10^4
60	2,6-$[(CH_3)_3C]_2$-4-NCC_6H_2OH	$HO_2^•$	−16.6	23.8	5.9×10^3	2.5×10^4
60	2,6-$[(CH_3)_3C]_2$-4-NCC_6H_2OH	*sec*-$RO_2^•$	− 13.1	25.4	3.3×10^3	1.5×10^4
60	2,6-$[(CH_3)_3C]_2$-4-NCC_6H_2OH	*tert*-$RO_2^•$	− 6.2	28.7	1.0×10^3	5.7×10^3
61	2,6-$[(CH_3)_3C]_2$-4-$H(O)CC_6H_2OH$	$HO_2^•$	−21.2	21.7	1.3×10^4	4.7×10^4
61	2,6-$[(CH_3)_3C]_2$-4-$H(O)CC_6H_2OH$	*sec*-$RO_2^•$	−17.7	23.3	7.1×10^3	2.9×10^4
61	2,6-$[(CH_3)_3C]_2$-4-$H(O)CC_6H_2OH$	*tert*-$RO_2^•$	−10.8	26.5	2.2×10^3	1.1×10^4
62	2,6-$[(CH_3)_3C]_2$-4-$CH_3OC_6H_2OH$	$HO_2^•$	−38.3	14.4	1.8×10^5	4.2×10^5
62	2,6-$[(CH_3)_3C]_2$-4-$CH_3OC_6H_2OH$	*sec*-$RO_2^•$	−34.8	15.8	1.1×10^5	2.8×10^5
62	2,6-$[(CH_3)_3C]_2$-4-$CH_3OC_6H_2OH$	*tert*-$RO_2^•$	−27.9	18.7	3.7×10^4	1.2×10^5
64	2,6-$[(CH_3)_3C]_2$-4-$CH_3C_6H_2OH$	$HO_2^•$	−30.0	17.8	5.2×10^4	1.5×10^5
64	2,6-$[(CH_3)_3C]_2$-4-$CH_3C_6H_2OH$	*sec*-$RO_2^•$	−26.5	19.3	3.0×10^4	9.7×10^4
64	2,6-$[(CH_3)_3C]_2$-4-$CH_3C_6H_2OH$	*tert*-$RO_2^•$	−19.6	22.4	9.8×10^3	3.8×10^4
68	2,6-$[(CH_3)_3C]_2$-4-$NO_2C_6H_2OH$	$HO_2^•$	−11.0	26.4	2.3×10^3	1.1×10^4
68	2,6-$[(CH_3)_3C]_2$-4-$NO_2C_6H_2OH$	*sec*-$RO_2^•$	−7.5	28.1	1.3×10^3	6.8×10^3
68	2,6-$[(CH_3)_3C]_2$-4-$NO_2C_6H_2OH$	*tert*-$RO_2^•$	−0.6	31.4	3.8×10^2	2.5×10^3
70	2,6-$[(CH_3)_3C]_2$-4-$C_6H_5C_6H_2OH$	$HO_2^•$	−31.3	17.3	6.2×10^4	1.8×10^5
70	2,6-$[(CH_3)_3C]_2$-4-$C_6H_5C_6H_2OH$	*sec*-$RO_2^•$	−27.8	18.8	3.6×10^4	1.1×10^5
70	2,6-$[(CH_3)_3C]_2$-4-$C_6H_5C_6H_2OH$	*tert*-$RO_2^•$	−20.9	21.8	1.2×10^4	4.6×10^4
73	2,6-$[(CH_3)_3C]_2$-4-$C_6H_5SC_6H_2OH$	$HO_2^•$	−22.6	21.1	1.6×10^4	5.6×10^4
73	2,6-$[(CH_3)_3C]_2$-4-$C_6H_5SC_6H_2OH$	*sec*-$RO_2^•$	−19.1	22.6	9.1×10^3	3.6×10^4
73	2,6-$[(CH_3)_3C]_2$-4-$C_6H_5SC_6H_2OH$	*tert*-$RO_2^•$	−12.2	25.8	2.9×10^3	1.4×10^4
89	2,6-$(C_6H_5)_2$-4-$NH_2CH_2C_6H_2OH$	$HO_2^•$	−46.1	11.3	5.4×10^5	1.1×10^6
89	2,6-$(C_6H_5)_2$-4-$NH_2CH_2C_6H_2OH$	*sec*-$RO_2^•$	−42.6	12.6	3.4×10^5	7.2×10^5
89	2,6-$(C_6H_5)_2$-4-$NH_2CH_2C_6H_2OH$	*tert*-$RO_2^•$	−35.7	15.4	1.2×10^5	3.1×10^5

No.	Phenol	$RO_2^•$	ΔH/ kJ mol^{-1}	E/ kJ mol^{-1}	k(333 K)/ l mol^{-1} s^{-1}	k(400 K)/ l mol^{-1} s^{-1}
91	$2,6\text{-}(C_6H_5)_2\text{-}4\text{-}HOOCCH_2C_6H_2OH$	$HO_2^•$	−43.1	12.4	3.6×10^5	7.7×10^5
91	$2,6\text{-}(C_6H_5)_2\text{-}4\text{-}HOOCCH_2C_6H_2OH$	*sec*-$RO_2^•$	−39.6	13.8	2.2×10^5	5.0×10^5
91	$2,6\text{-}(C_6H_5)_2\text{-}4\text{-}HOOCCH_2C_6H_2OH$	*tert*-$RO_2^•$	−32.7	16.8	7.4×10^4	2.0×10^5
94	$2,6\text{-}(C_6H_5)_2\text{-}4\text{-}CH_3SC_6H_2OH$	$HO_2^•$	−28.6	18.4	4.2×10^4	1.3×10^5
94	$2,6\text{-}(C_6H_5)_2\text{-}4\text{-}CH_3SC_6H_2OH$	*sec*-$RO_2^•$	−25.1	19.9	2.4×10^4	8.1×10^4
94	$2,6\text{-}(C_6H_5)_2\text{-}4\text{-}CH_3SC_6H_2OH$	*tert*-$RO_2^•$	−18.2	23.0	7.9×10^3	3.2×10^4
95	$2,6\text{-}(C_6H_5)_2\text{-}4\text{-}CH_3OC_6H_2OH$	$HO_2^•$	−41.0	13.3	2.6×10^5	5.9×10^5
95	$2,6\text{-}(C_6H_5)_2\text{-}4\text{-}CH_3OC_6H_2OH$	*sec*-$RO_2^•$	−37.5	14.7	1.6×10^5	3.8×10^5
95	$2,6\text{-}(C_6H_5)_2\text{-}4\text{-}CH_3OC_6H_2OH$	*tert*-$RO_2^•$	−30.6	17.6	5.6×10^4	1.6×10^5
122	$Si[2,6\text{-}[(CH_3)_3C]_2\text{-}4\text{-}CH_2CH_2OC_6H_2OH]_4$	$HO_2^•$	−28.1	18.6	1.6×10^5	4.8×10^5
122	$Si[2,6\text{-}[(CH_3)_3C]_2\text{-}4\text{-}CH_2CH_2OC_6H_2OH]_4$	*sec*-$RO_2^•$	−24.6	20.2	8.8×10^4	3.0×10^5
122	$Si[2,6\text{-}[(CH_3)_3C]_2\text{-}4\text{-}CH_2CH_2OC_6H_2OH]_4$	*tert*-$RO_2^•$	−17.7	23.3	2.8×10^4	1.2×10^5
123	$4,4'\text{-}CH_2SCH_2\text{-}[2,6\text{-}[(CH_3)_3C]_2C_6H_2OH$	$HO_2^•$	−29.1	18.2	9.0×10^4	2.6×10^5
123	$4,4'\text{-}CH_2SCH_2\text{-}[2,6\text{-}[(CH_3)_3C]_2C_6H_2OH$	*sec*-$RO_2^•$	−25.6	19.7	5.2×10^4	1.7×10^5
123	$4,4'\text{-}CH_2SCH_2\text{-}[2,6\text{-}[(CH_3)_3C]_2C_6H_2OH$	*tert*-$RO_2^•$	−18.7	22.8	1.7×10^4	7.2×10^4
124	$2,2'\text{-}SS\text{-}[4\text{-}CH_3\text{-}6\text{-}CH_3(C_6H_5)CHC_6H_2OH]_2$	$HO_2^•$	−33.8	16.2	9.2×10^4	2.5×10^5
124	$2,2'\text{-}SS\text{-}[4\text{-}CH_3\text{-}6\text{-}CH_3(C_6H_5)CHC_6H_2OH]_2$	*sec*-$RO_2^•$	−30.3	17.7	5.4×10^4	1.6×10^5
124	$2,2'\text{-}SS\text{-}[4\text{-}CH_3\text{-}6\text{-}CH_3(C_6H_5)CHC_6H_2OH]_2$	*tert*-$RO_2^•$	−23.4	20.7	1.8×10^4	6.3×10^4
126	Indophenol	$HO_2^•$	−42.0	12.9	3.0×10^4	6.6×10^5
126	Indophenol	*sec*-$RO_2^•$	−38.5	14.3	1.8×10^5	4.3×10^5
126	Indophenol	*tert*-$RO_2^•$	−31.6	17.1	6.6×10^4	1.9×10^5

Table 3.4
Enthalpies, activation energies and rate constants of reactions of phenoxyl radicals ($Ar_1O^•$) with secondary and tertiary hydroperoxides in hydrocarbon solutions calculated by formulas 1.15–1.17 and 1.21. The values of A, br_e and α, see Table 1.6

No. of ArOH	Phenoxyl	ROOH	ΔH/ kJ mol^{-1}	E/ kJ mol^{-1}	k (333 K)/ l mol^{-1} s^{-1}	k (400 K)/ l mol^{-1} s^{-1}
3	$2\text{-}HOC_6H_4O^•$	*sec*-ROOH	27.5	40.0	17	1.9×10^2
3	$2\text{-}HOC_6H_4O^•$	*tert*-ROOH	20.6	36.1	70	6.2×10^2
4	$2\text{-}HO\text{-}4\text{-}(CH_3)_3CC_6H_3O^•$	*sec*-ROOH	23.1	37.5	42	4.1×10^2
4	$2\text{-}HO\text{-}4\text{-}(CH_3)_3CC_6H_3O^•$	*tert*-ROOH	16.2	33.7	1.7×10^2	1.3×10^3
9	$4\text{-}CH_3C_6H_4O^•$	*sec*-ROOH	7.7	29.1	8.7×10^2	5.1×10^3
9	$4\text{-}CH_3C_6H_4O^•$	*tert*-ROOH	0.8	25.6	3.1×10^3	1.5×10^4
10	$(CH_3)_3C_6(O^•)(CH_2)_3O$	*sec*-ROOH	36.7	45.4	2.4	38
10	$(CH_3)_3C_6(O^•)(CH_2)_3O$	*tert*-ROOH	29.8	41.3	11	1.3×10^2
20	$(CH_3)_3C_6(O^•)CH_2CH(CH_3)O$	*sec*-ROOH	41.1	48.1	0.9	17
20	$(CH_3)_3C_6(O^•)CH_2CH(CH_3)O$	*tert*-ROOH	34.2	43.9	4.2	59
25	$1\text{-}C_{13}H_9O^•$	*sec*-ROOH	29.2	41.0	12	1.4×10^2
25	$1\text{-}C_{13}H_9O^•$	*tert*-ROOH	22.3	37.0	50	4.7×10^2
29	$1\text{-}C_{10}H_7O^•$	*sec*-ROOH	24.1	38.0	35	3.5×10^2
29	$1\text{-}C_{10}H_7O^•$	*tert*-ROOH	17.2	34.2	1.4×10^2	1.1×10^3
30	$2\text{-}C_{10}H_7O^•$	*sec*-ROOH	13.7	32.3	2.8×10^2	1.9×10^3
30	$2\text{-}C_{10}H_7O^•$	*tert*-ROOH	6.8	28.7	1.0×10^3	5.7×10^3
32	$1\text{-}C_{14}H_9O^•$	*sec*-ROOH	12.8	31.8	3.3×10^2	2.3×10^3
32	$1\text{-}C_{14}H_9O^•$	*tert*-ROOH	5.9	28.2	1.2×10^3	6.6×10^3
36	$C_6H_5O^•$	*sec*-ROOH	−1.5	24.4	4.8×10^3	2.1×10^4
36	$C_6H_5O^•$	*tert*-ROOH	−8.4	21.1	1.6×10^4	5.6×10^4
38	$4\text{-}CH_3COC_6H_4O^•$	*sec*-ROOH	−4.0	23.2	7.3×10^3	3.0×10^4
38	$4\text{-}CH_3COC_6H_4O^•$	*tert*-ROOH	−10.9	19.9	2.4×10^4	8.1×10^4
40	$4\text{-}NH_2C_6H_4O^•$	*sec*-ROOH	10.8	30.8	4.7×10^2	3.0×10^3
40	$4\text{-}NH_2C_6H_4O^•$	*tert*-ROOH	3.9	27.2	1.7×10^3	9.0×10^3

No. of ArOH	Phenoxyl	ROOH	ΔH/ kJ mol^{-1}	E/ kJ mol^{-1}	k (333 K)/ l mol^{-1} s^{-1}	k (400 K)/ l mol^{-1} s^{-1}
41	$4\text{-}C_6H_5CH_2OC_6H_4O^{\bullet}$	*sec*-ROOH	18.0	34.6	1.2×10^2	9.7×10^2
41	$4\text{-}C_6H_5CH_2OC_6H_4O^{\bullet}$	*tert*-ROOH	11.1	30.9	4.6×10^2	3.0×10^3
44	$4\text{-}CH_3(CH_2)_3OC_6H_4O^{\bullet}$	*sec*-ROOH	19.7	35.6	83	7.2×10^2
44	$4\text{-}CH_3(CH_2)_3OC_6H_4O^{\bullet}$	*tert*-ROOH	12.8	31.8	3.3×10^2	2.3×10^3
46	$4\text{-}(CH_3)_3CC_6H_4O^{\bullet}$	*sec*-ROOH	8.6	29.6	7.3×10^2	4.4×10^3
46	$4\text{-}(CH_3)_3CC_6H_4O^{\bullet}$	*tert*-ROOH	1.7	26.0	2.7×10^3	1.3×10^4
47	$2,4\text{-}[(CH_3)_3C]_2C_6H_3O^{\bullet}$	*sec*-ROOH	8.0	29.3	8.1×10^2	4.8×10^3
47	$2,4\text{-}[(CH_3)_3C]_2C_6H_3O^{\bullet}$	*tert*-ROOH	1.1	25.7	3.0×10^3	1.4×10^4
74	$2\text{-}(CH_3)_3C\text{-}4,6\text{-}(CH_3)_2C_6H_2O^{\bullet}$	*sec*-ROOH	11.7	31.2	4.1×10^2	2.7×10^3
74	$2\text{-}(CH_3)_3C\text{-}4,6\text{-}(CH_3)_2C_6H_2O^{\bullet}$	*tert*-ROOH	4.8	27.6	1.5×10^3	8.0×10^3
75	$2\text{-}(CH_3)_3C\text{-}4\text{-}CH_3OC_6H_3O^{\bullet}$	*sec*-ROOH	24.1	38.0	35	3.5×10^2
75	$2\text{-}(CH_3)_3C\text{-}4\text{-}CH_3OC_6H_3O^{\bullet}$	*tert*-ROOH	17.2	34.2	1.4×10^2	1.1×10^3
78	$4\text{-}CH_3OC_6H_4O^{\bullet}$	*sec*-ROOH	16.5	33.8	1.6×10^2	1.2×10^3
78	$4\text{-}CH_3OC_6H_4O^{\bullet}$	*tert*-ROOH	9.6	30.1	6.1×10^2	3.8×10^3
79	$2,3\text{-}(CH_3)_2C_6H_3O^{\bullet}$	*sec*-ROOH	12.0	31.4	3.8×10^2	2.5×10^3
79	$2,3\text{-}(CH_3)_2C_6H_3O^{\bullet}$	*tert*-ROOH	5.1	27.8	1.4×10^3	7.5×10^3
80	$3,4\text{-}(CH_3)_2C_6H_3O^{\bullet}$	*sec*-ROOH	12.9	31.9	3.2×10^2	2.2×10^3
80	$3,4\text{-}(CH_3)_2C_6H_3O^{\bullet}$	*tert*-ROOH	6.0	28.2	1.2×10^3	6.6×10^3
81	$3,5\text{-}(CH_3)_2C_6H_3O^{\bullet}$	*sec*-ROOH	11.5	31.1	4.2×10^2	2.8×10^3
81	$3,5\text{-}(CH_3)_2C_6H_3O^{\bullet}$	*tert*-ROOH	4.6	27.5	1.6×10^3	8.2×10^3
82	$2,4\text{-}(CH_3)_2C_6H_3O^{\bullet}$	*sec*-ROOH	7.0	28.8	9.7×10^2	5.6×10^3
82	$2,4\text{-}(CH_3)_2C_6H_3O^{\bullet}$	*tert*-ROOH	0.1	25.3	3.4×10^3	1.6×10^4
83	$2,6\text{-}(CH_3)_2C_6H_3O^{\bullet}$	*sec*-ROOH	10.6	30.6	5.1×10^2	3.2×10^3
83	$2,6\text{-}(CH_3)_2C_6H_3O^{\bullet}$	*tert*-ROOH	3.7	27.0	1.9×10^3	9.6×10^3
85	$4\text{-}(CH_3)_2NC_6H_4O^{\bullet}$	*sec*-ROOH	38.7	46.6	1.6	26
85	$4\text{-}(CH_3)_2NC_6H_4O^{\bullet}$	*tert*-ROOH	31.8	42.5	6.9	90
110	$2,4,6\text{-}(CH_3)_3C_6H_2O^{\bullet}$	*sec*-ROOH	18.7	35.0	1.0×10^2	8.6×10^2
110	$2,4,6\text{-}(CH_3)_3C_6H_2O^{\bullet}$	*tert*-ROOH	11.8	31.3	3.9×10^2	2.6×10^3

No. of ArOH	Phenoxyl	ROOH	ΔH/ kJ mol^{-1}	E/ kJ mol^{-1}	k (333 K)/ l mol^{-1} s^{-1}	k (400 K)/ l mol^{-1} s^{-1}
113	2,6-$(CH_3)_2$-4-$COOHCH_2$-C_6H_2OH	*sec*-ROOH	17.5	34.4	1.3×10^2	1.0×10^3
113	2,6-$(CH_3)_2$-4-$COOHCH_2$-C_6H_2OH	*tert*-ROOH	10.6	30.6	5.1×10^2	3.2×10^3
121	3-HOC_6H_4OH	*sec*-ROOH	–0.3	25.0	7.6×10^3	3.4×10^4
121	3-HOC_6H_4OH	*tert*-ROOH	–7.2	21.7	2.6×10^4	9.4×10^4
130	α-Tocopheroxyl	*sec*-ROOH	37.8	46.0	1.9	31
130	α-Tocopheroxyl	*tert*-ROOH	30.9	41.9	8.6	1.1×10^2
131	β-Tocopheroxyl	*sec*-ROOH	33.1	43.3	5.2	71
131	β-Tocopheroxyl	*tert*-ROOH	26.2	39.2	23	2.4×10^2
133	δ-Tocopheroxyl	*sec*-ROOH	48.5	52.7	0.2	4.2
133	δ-Tocopheroxyl	*tert*-ROOH	41.6	48.4	0.8	15

Table 3.5

Enthalpies, activation energies and rate constants of reaction of sterically hindered phenoxyls ($Ar_2O^•$) with secondary and tertiary hydroperoxides in hydrocarbon solutions calculated by formulas 1.15–1.17 and 1.21. The values of A, br_e and α, see Table 1.6

No. of ArOH	Phenoxyl	ROOH	ΔH/ kJ mol^{-1}	E/ kJ mol^{-1}	k (400 K)/ l mol^{-1} s^{-1}	k (500 K)/ l mol^{-1} s^{-1}
1	1-[3′,5′-$[(CH_3)_2C]_2$-4′-($^•O$-$C_6H_2CH_2$)]-3,5-[$[(CH_3)_2C]_2$-4′-$HOC_6H_2CH_2$]$_2$-C_6H_3	*sec*-ROOH	20.6	42.6	88	1.1×10^3
1	1-[3′,5′-$[(CH_3)_2C]_2$-4′-($^•O$-$C_6H_2CH_2$)]-3,5-[$[(CH_3)_2C]_2$-4′-$HOC_6H_2CH_2$]$_2$-C_6H_3	*tert*-ROOH	13.7	38.8	2.7×10^2	2.8×10^3
24	[4,6-$[(CH_3)_3C]_2C_6H_2O^•$]-CH_2-2-[4,6-$[(CH_3)_3C]_2C_6H_2OH$	*sec*-ROOH	24.6	44.8	45	6.7×10^2
24	[4,6-$[(CH_3)_3C]_2C_6H_2O^•$]-CH_2-2-[4,6-$[(CH_3)_3C]_2C_6H_2OH$	*tert*-ROOH	17.7	41.0	1.4×10^2	1.7×10^3
31	C-[4-$(CH_2)_2COOCH_2$-2,6-$[(CH_3)_3C]_2C_6H_2O^•$] [4-$(CH_2)_2COOCH_2$-2,6-$[(CH_3)C]_2C_6H_2OH$]$_3$	*sec*-ROOH	24.5	44.7	47	6.8×10^2

No. of ArOH	Phenoxyl	ROOH	ΔH/ kJ mol^{-1}	E/ kJ mol^{-1}	k (400 K)/ l mol^{-1} s^{-1}	k (500 K)/ l mol^{-1} s^{-1}
31	$C-[4-(CH_2)_2COOCH_2-2,6-[(CH_3)_3C]_2C_6H_2O^\bullet]$ $[4-(CH_2)_2COOCH_2-2,6-[(CH_3)C]_2C_6H_2OH]_3$	*tert*-ROOH	17.6	40.9	1.5×10^2	1.7×10^3
49	$2,4-[(CH_3)_3C]_2-6-CH_3-C_6H_2O^\bullet$	*sec*-ROOH	11.6	37.7	3.8×10^2	3.7×10^3
49	$2,4-[(CH_3)_3C]_2-6-CH_3-C_6H_2O^\bullet$	*tert*-ROOH	4.7	34.1	1.1×10^3	8.8×10^3
50	$[2,6-[(CH_3)_3C]_2C_6H_2O^\bullet]-$ $4'-[2,6-[(CH_3)_3C]_2C_6H_2OH]$	*sec*-ROOH	22.9	43.8	61	8.5×10^2
50	$[2,6-[(CH_3)_3C]_2C_6H_2O^\bullet]-$ $4'-[2,6-[(CH_3)_3C]_2C_6H_2OH]$	*tert*-ROOH	16.0	40.0	1.9×10^2	2.1×10^3
51	$2,6-[(CH_3)_3C]_2C_6H_3O^\bullet$	*sec*-ROOH	18.8	41.6	1.2×10^2	1.4×10^3
51	$2,6-[(CH_3)_3C]_2C_6H_3O^\bullet$	*tert*-ROOH	11.9	37.9	3.6×10^2	3.5×10^3
52	$2,6-[(CH_3)_3C]_2-4-CH_3CONHC_6H_2O^\bullet$	*sec*-ROOH	38.8	53.0	3.8	93
52	$2,6-[(CH_3)_3C]_2-4-CH_3CONHC_6H_2O^\bullet$	*tert*-ROOH	31.9	48.9	13	2.5×10^2
53	$2,6-[(CH_3)_3C]_2-4-CH_3COC_6H_2O^\bullet$	*sec*-ROOH	20.0	42.2	99	1.2×10^3
53	$2,6-[(CH_3)_3C]_2-4-CH_3COC_6H_2O^\bullet$	*tert*-ROOH	13.1	38.5	3.0×10^2	3.0×10^3
54	$2,6-[(CH_3)_3C]_2-4-C_6H_5CH_2C_6H_2O^\bullet$	*sec*-ROOH	25.8	45.4	38	5.8×10^2
54	$2,6-[(CH_3)_3C]_2-4-C_6H_5CH_2C_6H_2O^\bullet$	*tert*-ROOH	18.9	41.6	1.2×10^2	1.4×10^3
55	$2,6-[(CH_3)_3C]_2-4-(CH_3)_3COC_6H_2O^\bullet$	*sec*-ROOH	34.2	50.3	8.6	1.8×10^2
55	$2,6-[(CH_3)_3C]_2-4-(CH_3)_3COC_6H_2O^\bullet$	*tert*-ROOH	27.3	46.3	29	4.7×10^2
56	$2,6-[(CH_3)_3C]_2-4-(CH_3)_3COC(O)C_6H_2O^\bullet$	*sec*-ROOH	17.8	41.0	1.4×10^2	1.7×10^3
56	$2,6-[(CH_3)_3C]_2-4-(CH_3)_3COC(O)C_6H_2O^\bullet$	*tert*-ROOH	10.9	37.3	4.3×10^2	4.1×10^3
57	$2,6-[(CH_3)_3C]_2-4-HOOCC_6H_2O^\bullet$	*sec*-ROOH	16.7	40.4	1.7×10^2	1.9×10^3
57	$2,6-[(CH_3)_3C]_2-4-HOOCC_6H_2O^\bullet$	*tert*-ROOH	9.8	36.8	5.0×10^2	4.6×10^3
58	$2,6-[(CH_3)_3C]_2-4-ClC_6H_2O^\bullet$	*sec*-ROOH	21.0	42.8	82	1.1×10^3
58	$2,6-[(CH_3)_3C]_2-4-ClC_6H_2O^\bullet$	*tert*-ROOH	14.1	39.0	2.6×10^3	2.7×10^3
60	$2,6-[(CH_3)_3C]_2-4-NCC_6H_2O^\bullet$	*sec*-ROOH	13.1	38.5	3.0×10^2	3.0×10^3
60	$2,6-[(CH_3)_3C]_2-4-NCC_6H_2O^\bullet$	*tert*-ROOH	6.2	34.9	8.9×10^2	7.2×10^3
61	$2,6-[(CH_3)_3C]_2-4-H(O)CC_6H_2O^\bullet$	*sec*-ROOH	17.7	41.0	1.4×10^2	1.7×10^3
61	$2,6-[(CH_3)_3C]_2-4-H(O)CC_6H_2O^\bullet$	*tert*-ROOH	10.8	37.3	4.3×10^2	4.1×10^3

No. of ArOH	Phenoxyl	ROOH	ΔH/ kJ mol^{-1}	E/ kJ mol^{-1}	k (400 K)/ l mol^{-1} s^{-1}	k (500 K)/ l mol^{-1} s^{-1}
62	$2,6\text{-}[(CH_3)_3C]_2\text{-}4\text{-}CH_3OC_6H_2O^{\bullet}$	*sec*-ROOH	34.8	50.6	7.9	1.7×10^2
62	$2,6\text{-}[(CH_3)_3C]_2\text{-}4\text{-}CH_3OC_6H_2O^{\bullet}$	*tert*-ROOH	27.9	46.6	26	4.3×10^2
64	$2,6\text{-}[(CH_3)_3C]_2\text{-}4\text{-}CH_3C_6H_2O^{\bullet}$	*sec*-ROOH	26.5	45.8	33	5.2×10^2
64	$2,6\text{-}[(CH_3)_3C]_2\text{-}4\text{-}CH_3C_6H_2O^{\bullet}$	*tert*-ROOH	19.6	42.0	1.0×10^2	1.3×10^3
68	$2,6\text{-}[(CH_3)_3C]_2\text{-}4\text{-}NO_2C_6H_2O^{\bullet}$	*sec*-ROOH	7.5	35.6	7.2×10^2	6.1×10^3
68	$2,6\text{-}[(CH_3)_3C]_2\text{-}4\text{-}NO_2C_6H_2O^{\bullet}$	*tert*-ROOH	0.6	32.0	2.1×10^3	1.5×10^4
70	$2,6\text{-}[(CH_3)_3C]_2\text{-}4\text{-}C_6H_5C_6H_2O^{\bullet}$	*sec*-ROOH	27.8	46.6	26	4.3×10^2
70	$2,6\text{-}[(CH_3)_3C]_2\text{-}4\text{-}C_6H_5C_6H_2O^{\bullet}$	*tert*-ROOH	20.9	42.7	85	1.1×10^3
73	$2,6\text{-}[(CH_3)_3C]_2\text{-}4\text{-}C_6H_5SC_6H_2O^{\bullet}$	*sec*-ROOH	19.1	41.7	1.1×10^2	1.4×10^3
73	$2,6\text{-}[(CH_3)_3C]_2\text{-}4\text{-}C_6H_5SC_6H_2O^{\bullet}$	*tert*-ROOH	12.2	38.0	3.5×10^2	3.4×10^3
89	$2,6\text{-}(C_6H_5)_2\text{-}4\text{-}NH_2CH_2C_6H_2O^{\bullet}$	*sec*-ROOH	42.6	55.2	2.0	55
89	$2,6\text{-}(C_6H_5)_2\text{-}4\text{-}NH_2CH_2C_6H_2O^{\bullet}$	*tert*-ROOH	35.7	51.1	6.8	1.5×10^2
90	$2,6\text{-}(C_6H_5)_2\text{-}4\text{-}CNCH_2C_6H_2O^{\bullet}$	*sec*-ROOH	54.4	62.6	0.21	9.2
90	$2,6\text{-}(C_6H_5)_2\text{-}4\text{-}CNCH_2C_6H_2O^{\bullet}$	*tert*-ROOH	47.5	58.2	0.80	27
91	$2,6\text{-}(C_6H_5)_2\text{-}4\text{-}HOOCCH_2C_6H_2O^{\bullet}$	*sec*-ROOH	39.6	53.4	3.4	84
91	$2,6\text{-}(C_6H_5)_2\text{-}4\text{-}HOOCCH_2C_6H_2O^{\bullet}$	*tert*-ROOH	32.7	49.5	11	2.2×10^2
92	$2,6\text{-}(C_6H_5)_2\text{-}4\text{-}HSC_6H_2O^{\bullet}$	*sec*-ROOH	25.5	45.3	39	5.9×10^2
92	$2,6\text{-}(C_6H_5)_2\text{-}4\text{-}HSC_6H_2O^{\bullet}$	*tert*-ROOH	18.6	41.5	1.2×10^2	1.5×10^3
94	$2,6\text{-}(C_6H_5)_2\text{-}4\text{-}CH_3SC_6H_2O^{\bullet}$	*sec*-ROOH	25.1	45.0	43	6.4×10^2
94	$2,6\text{-}(C_6H_5)_2\text{-}4\text{-}CH_3SC_6H_2O^{\bullet}$	*tert*-ROOH	18.2	41.2	1.3×10^2	1.6×10^3
95	$2,6\text{-}(C_6H_5)_2\text{-}4\text{-}CH_3OC_6H_2O^{\bullet}$	*sec*-ROOH	37.5	52.2	4.9	1.1×10^2
95	$2,6\text{-}(C_6H_5)_2\text{-}4\text{-}CH_3OC_6H_2O^{\bullet}$	*tert*-ROOH	30.6	48.2	16	2.9×10^2
96	$2,6\text{-}(C_6H_5)_2\text{-}4\text{-}C_{18}H_{37}C_6H_2O^{\bullet}$	*sec*-ROOH	36.9	51.8	5.5	1.2×10^2
96	$2,6\text{-}(C_6H_5)_2\text{-}4\text{-}C_{18}H_{37}C_6H_2O^{\bullet}$	*tert*-ROOH	30.0	47.8	18	3.2×10^2
97	$2,6\text{-}(C_6H_5)_2\text{-}4\text{-}C_{24}H_{49}OC_6H_2O^{\bullet}$	*sec*-ROOH	36.3	51.5	6.0	1.3×10^2
97	$2,6\text{-}(C_6H_5)_2\text{-}4\text{-}C_{24}H_{49}OC_6H_2O^{\bullet}$	*tert*-ROOH	29.4	47.5	20	3.5×10^2
122	$Si[2,6\text{-}[(CH_3)_3C]_2\text{-}4\text{-}OCH_2CH_2C_6H_2OH]_3$ $[[(CH_3)_3C]_2\text{-}4\text{-}CH_2CH_2OC_6H_2O^{\bullet}]$	*sec*-ROOH	24.6	44.8	45	6.7×10^2

No. of ArOH	Phenoxyl	ROOH	ΔH/ kJ mol^{-1}	E/ kJ mol^{-1}	k (400 K)/ l mol^{-1} s^{-1}	k (500 K)/ l mol^{-1} s^{-1}
122	$Si[2,6-[(CH_3)_3C]_2-4-OCH_2CH_2C_6H_2OH]_3$ $[[(CH_3)_3C]_2-4-CH_2CH_2OC_6H_2O^{\bullet}]$	*tert*-ROOH	17.7	41.0	1.4×10^2	1.7×10^3
123	$4-[2,6-[(CH_3)_3C]_2C_6H_2OH]-CH_2SCH_2-$ $4'-[2,6-[(CH_3)_3C]C_6H_2O^{\bullet}]$	*sec*-ROOH	25.6	45.3	39	5.9×10^2
123	$4-[2,6-[(CH_3)_3C]_2C_6H_2OH]-CH_2SCH_2-$ $4'-[2,6-[(CH_3)_3C]C_6H_2O^{\bullet}]$	*tert*-ROOH	18.7	41.5	1.2×10^2	1.5×10^3

Table 3.6
Enthalpies, activation energies and rate constants of reactions of phenoxyls ($Ar_1O^{\bullet}$) with cumene calculated by formulas 1.15–1.17 and 1.20. The values of A, br_e and α, see Table 1.6

No. of ArOH	Phenoxyl	ΔH/ kJ mol^{-1}	E/ kJ mol^{-1}	k (333 K)/ l mol^{-1} s^{-1}	k (400 K)/ l mol^{-1} s^{-1}
3	$2-HOC_6H_4O^{\bullet}$	11.0	61.4	2.3×10^{-2}	0.96
4	$2-HO-4-(CH_3)_3CC_6H_3O^{\bullet}$	6.6	59.4	4.8×10^{-2}	1.8
9	$4-CH_3C_6H_4O^{\bullet}$	−8.8	52.6	0.56	13.5
10	$(CH_3)_3C_6(O^{\bullet})(CH_2)_3O$	20.2	65.6	5.1×10^{-3}	0.27
20	$(CH_3)_3C_6(O^{\bullet})CH_2CH(CH_3)O$	24.6	67.7	2.4×10^{-3}	0.14
25	$1-C_{13}H_9O^{\bullet}$	12.7	62.2	1.8×10^{-2}	0.75
29	$1-C_{10}H_7O^{\bullet}$	7.6	59.8	4.2×10^{-2}	1.5
30	$2-C_{10}H_7O^{\bullet}$	−2.8	55.2	0.22	6.2
32	$1-C_{14}H_9O^{\bullet}$	−3.7	54.8	0.25	7.0
36	$C_6H_5O^{\bullet}$	−18.0	48.8	2.2	42
38	$4-CH_3COC_6H_4O^{\bullet}$	−20.5	47.8	3.2	57
40	$4-NH_2C_6H_4O^{\bullet}$	−5.7	54.0	0.34	8.9
41	$4-C_6H_5CH_2OC_6H_4O^{\bullet}$	1.5	57.1	0.11	3.5
44	$4-CH_3(CH_2)_3OC_6H_4O^{\bullet}$	3.2	57.9	8.3×10^{-2}	2.7
46	$4-(CH_3)_3CC_6H_4O^{\bullet}$	−7.9	53.0	0.49	12
47	$2,4-[(CH_3)_3C]_2C_6H_3O^{\bullet}$	−8.5	52.8	0.52	13
74	$2-(CH_3)_3C-4,6-(CH_3)_2C_6H_2O^{\bullet}$	−4.8	54.4	0.29	6.8

No. of ArOH	Phenoxyl	ΔH/ kJ mol^{-1}	E/ kJ mol^{-1}	k (333 K)/ l mol^{-1} s^{-1}	k (400 K)/ l mol^{-1} s^{-1}
75	2-$(CH_3)_3C$-4-$CH_3OC_6H_3O^{\bullet}$	7.6	59.8	4.2×10^{-2}	1.6
78	4-$CH_3OC_6H_4O^{\bullet}$	0.0	56.5	0.14	4.2
79	2,3-$(CH_3)_2C_6H_3O^{\bullet}$	−4.5	54.5	0.28	7.6
80	3,4-$(CH_3)_2C_6H_3O^{\bullet}$	−3.6	54.9	0.24	6.8
81	3,5-$(CH_3)_2C_6H_3O^{\bullet}$	−5.0	54.3	0.30	8.1
82	2,4-$(CH_3)_2C_6H_3O^{\bullet}$	−9.5	52.0	0.70	16
83	2,6-$(CH_3)_2C_6H_3O^{\bullet}$	−5.9	53.9	0.35	9.2
85	4-$(CH_3)_2NC_6H_4O^{\bullet}$	22.2	66.6	3.6×10^{-3}	0.20
87	4-$NO_2C_6H_4O$	−20.7	47.7	3.3	59
88	4-$C_6H_5C_6H_4O$	−8.4	52.8	0.52	13
101	4-$HO(O)CC_6H_4O$	−20.7	47.7	3.3	59
102	4-ClC_6H_4O	−19.6	48.2	2.8	51
106	4-CNC_6H_4O	−19.2	48.4	2.6.	48
108	2-CH_3-4-$(CH_3)_3CC_6H_3O$	−8.9	52.6	0.56	14
109	3,4,5-$(CH_3)_3CC_6H_2O$	−5.8	54.0	0.34	8.9
110	2,4,6-$(CH_3)_3C_6H_2O^{\bullet}$	2.2	57.4	9.9×10^{-2}	3.2
111	2-CH_3-4-NH_2CH_2-6-$(CH_3)_3CC_6H_2O$	0.4	56.7	0.13	3.9
112	2,6-$(CH_3)_2$4-$NH_2CH_2C_6H_2O$	2.9	57.7	8.9×10^{-2}	2.9
116	2,3,5,6-$(CH_3)_4$-C_6HO	−0.2	56.4	0.14	4.3
117	$(CH_3)_5C_6HO$	10.5	61.2	2.5×10^{-2}	1.0
118	2,3,5,6-$(CH_3)_4$-4-CH_3OC_6O	10.9	61.4	2.3×10^{-2}	0.96
130	α-Tocopheroxyl	21.3	66.2	4.1×10^{-3}	0.23
131	β-Tocopheroxyl	16.6	64.0	9.1×10^{-3}	0.44
133	δ-Tocopheroxyl	32.0	71.3	6.5×10^{-4}	0.49

Table 3.7
Enthalpies, activation energies and rate constants of reaction of *para*-methoxyphenoxyl radical ($Ar_1O^•$) with sterically hindered phenols (Ar_2OH) in hydrocarbon solutions calculated by formulas 1.15–1.17 and 1.21. The values of A, br_e and α, see Table 1.6

No.	Phenol	ΔH/ kJ mol^{-1}	E/ kJ mol^{-1}	k (333 K)/ l mol^{-1} s^{-1}	k (400 K)/ l mol^{-1} s^{-1}
1	1,3,5-[3′,5′-[$(CH_3)_3C]_2$-4′-HO-$C_6H_5CH_2]_3C_6H_3$	−4.1	22.2	9.9×10^4	3.9×10^5
24	2,2′-CH_2-[4,6-[$(CH_3)_3C]_2C_6H_2OH]_2$	−8.1	20.2	1.4×10^5	4.6×10^5
31	C-[4-$(CH_2)_2COOCH_2$-2,6-[$(CH_3)_3C]_2C_6H_2OH]_4$	−8.0	20.3	2.6×10^5	8.8×10^5
49	2,4-[$(CH_3)_3C]_2$-6-CH_3-C_6H_2OH	4.9	26.7	6.5×10^3	3.3×10^4
50	4,4′-[2,6-[$(CH_3)_3C]_2C_6H_2OH]_2$	−6.4	21.0	1.0×10^5	3.6×10^5
51	2,6-[$(CH_3)_3C]_2C_6H_3OH$	−2.3	23.0	2.5×10^4	9.9×10^4
52	2,6-[$(CH_3)_3C]_2$-4-$CH_3CONHC_6H_2OH$	−22.3	13.7	7.1×10^5	1.6×10^6
53	2,6-[$(CH_3)_3C]_2$-4-$CH_3COC_6H_2OH$	−3.5	22.4	3.1×10^4	1.2×10^5
54	2,6-[$(CH_3)_3C]_2$-4-$C_6H_5CH_2C_6H_2OH$	−9.3	19.6	8.4×10^4	2.8×10^5
55	2,6-[$(CH_3)_3C]_2$-4-$(CH_3)_3COC_6H_2OH$	−17.7	15.7	3.4×10^5	8.9×10^5
56	2,6-[$(CH_3)_3C]_2$-4-$(CH_3)_3COC(O)C_6H_2OH$	−1.3	23.5	2.1×10^4	8.5×10^4
57	2,6-[$(CH_3)_3C]_2$-4-$HOOCC_6H_2OH$	−0.2	24.3	1.5×10^4	6.7×10^4
58	2,6-[$(CH_3)_3C]_2$-4-ClC_6H_2OH	−4.5	22.0	3.5×10^4	1.3×10^5
60	2,6-[$(CH_3)_3C]_2$-4-NCC_6H_2OH	3.4	25.9	8.7×10^3	4.1×10^4
61	2,6-[$(CH_3)_3C]_2$-4-$H(O)CC_6H_2OH$	−2.2	23.1	2.4×10^4	9.6×10^4
62	2,6-[$(CH_3)_3C]_2$-4-$CH_3OC_6H_2OH$	−18.3	15.5	3.7×10^5	9.5×10^5
64	2,6-[$(CH_3)_3C]_2$-4-$CH_3C_6H_2OH$	−10.0	19.3	9.4×10^4	3.0×10^5
65	2,6-[$(CH_3)_3C]_2$-4-$NH_2CH_2C_6H_2OH$	−14.2	17.4	1.9×10^5	5.3×10^5
66	2,6-[$(CH_3)_3C]_2$-4-$HO(O)CCH_2C_6H_2OH$	−12.1	18.3	1.3×10^5	4.1×10^5
67	2,6-[$(CH_3)_3C]_2$-4-$(CH_3)_3CCH_2C_6H_2OH$	−9.2	19.7	8.1×10^4	2.7×10^5
68	2,6-[$(CH_3)_3C]_2$-4-$NO_2C_6H_2OH$	9.0	28.8	3.0×10^3	1.7×10^4
70	2,6-[$(CH_3)_3C]_2$-4-$C_6H_5C_6H_2OH$	−11.3	18.7	1.2×10^5	3.6×10^5
71	2,6-[$(CH_3)_3C]_2$-4-$(C_6H_5)_2CHC_6H_2OH$	−6.7	20.9	5.3×10^4	1.9×10^5
72	2,6-[$(CH_3)_3C]_2$-4-$C_{18}H_{37}OC(O)CH_2CH_2C_6H_2OH$	−9.4	19.6	8.4×10^4	2.8×10^5
73	2,6-[$(CH_3)_3C]_2$-4-$C_6H_5SC_6H_2OH$	−2.6	22.9	2.6×10^4	1.2×10^4
89	2,6-$(C_6H_5)_2$-4-$NH_2CH_2C_6H_2OH$	−26.1	12.0	1.3×10^6	2.7×10^6

No.	Phenol	ΔH/ kJ mol^{-1}	E/ kJ mol^{-1}	k (333 K)/ l mol^{-1} s^{-1}	k (400 K)/ l mol^{-1} s^{-1}
91	$2,6\text{-}(C_6H_5)_2\text{-}4\text{-}HOOCCH_2C_6H_2OH$	−23.1	13.3	8.2×10^5	1.8×10^6
92	$2,6\text{-}[CH(CH_3)C_6H_5]_2\text{-}4\text{-}HSC_6H_2OH$	−9.0	19.8	7.8×10^4	2.6×10^5
93	$2,6\text{-}[CH(CH_3)C_6H_5]_2\text{-}4\text{-}CH_3(CH_2)_3SC_6H_2OH$	−7.8	20.4	6.3×10^4	2.2×10^5
94	$2,6\text{-}(C_6H_5)_2\text{-}4\text{-}CH_3SC_6H_2OH$	−8.6	20.0	7.3×10^4	2.4×10^5
95	$2,6\text{-}(C_6H_5)_2\text{-}4\text{-}CH_3OC_6H_2OH$	−21.0	14.3	5.7×10^5	1.4×10^6
96	$2,6\text{-}(C_6H_5)_2\text{-}4\text{-}C_{18}H_{37}OC_6H_2OH$	−20.4	14.6	5.1×10^5	1.2×10^6
97	$2,6\text{-}(C_6H_5)_2\text{-}4\text{-}C_{24}H_{49}OC_6H_2OH$	−19.8	14.8	4.8×10^5	1.1×10^6
122	$Si[2,6\text{-}[(CH_3)_3C]_2\text{-}4\text{-}CH_2CH_2OC_6H_2OH]_4$	−8.1	20.2	2.7×10^5	9.2×10^5
123	$4,4'\text{-}CH_2SCH_2\text{-}[2,6\text{-}[(CH_3)_3C]_2C_6H_2OH]_2$	−9.1	19.7	1.6×10^5	2.7×10^5
126	Indophenol	−22.0	13.9	1.3×10^6	3.0×10^6

Table 3.8

Enthalpies, activation energies and rate constants of reaction of 2,4,6-*tert*-butylphenoxyl radical ($Ar_2O^•$) with different phenols (Ar_1OH) in hydrocarbon solutions calculated by formulas 1.15–1.17 and 1.21. The values of A, br_e and α, see Table 1.6

No.	Phenol	ΔH / kJ mol^{-1}	E / kJ mol^{-1}	k (333 K)/ l mol^{-1} s^{-1}	k (400 K)/ l mol^{-1} s^{-1}
3	$2\text{-}HOC_6H_4OH$	−1.0	23.7	1.9×10^4	8.0×10^4
4	$2\text{-}HO\text{-}4\text{-}(CH_3)_3CC_6H_3OH$	3.4	25.9	8.7×10^3	4.1×10^4
9	$4\text{-}CH_3C_6H_4OH$	18.8	34.1	4.2×10^2	3.5×10^3
10	$(CH_3)_3C_6(OH)(CH_2)_3O$	−10.2	19.2	9.7×10^4	3.1×10^5
20	$(CH_3)_3C_6(OH)CH_2CH(CH_3)O$	−4.6	17.2	2.0×10^5	5.7×10^5
25	$1\text{-}HOC_{13}H_9$	−2.7	22.8	2.7×10^4	1.1×10^5
29	$1\text{-}HOC_{10}H_7$	2.4	25.4	1.0×10^4	4.8×10^4
30	$2\text{-}HOC_{10}H_7$	12.8	30.8	1.5×10^3	9.5×10^3
32	$1\text{-}HOC_{14}H_9$	13.7	31.3	1.2×10^3	8.2×10^3
36	C_6H_5OH	28.0	39.3	68	7.4×10^2
38	$4\text{-}CH_3COC_6H_4OH$	30.5	40.8	40	4.7×10^2
40	$4\text{-}NH_2C_6H_4OH$	15.7	32.4	8.3×10^2	5.9×10^3
41	$4\text{-}C_6H_5CH_2OC_6H_4OH$	8.5	28.5	3.4×10^3	1.9×10^4

No.	Phenol	ΔH / kJ mol^{-1}	E / kJ mol^{-1}	k (333 K)/ l mol^{-1} s^{-1}	k (400 K)/ l mol^{-1} s^{-1}
44	$4\text{-}CH_3(CH_2)_3OC_6H_4OH$	6.8	27.6	4.7×10^3	2.5×10^4
46	$4\text{-}(CH_3)_3CC_6H_4OH$	17.9	33.6	5.4×10^2	4.1×10^3
47	$2,4\text{-}[(CH_3)_3C]_2C_6H_3OH$	18.5	33.9	5.4×10^2	3.7×10^3
74	$2\text{-}(CH_3)_3C\text{-}4,6\text{-}(CH_3)_2C_6H_2OH$	14.8	31.9	9.9×10^2	6.8×10^3
75	$2\text{-}(CH_3)_3C\text{-}4\text{-}CH_3OC_6H_3OH$	2.4	25.4	1.0×10^4	4.8×10^4
78	$4\text{-}CH_3OC_6H_4OH$	10.0	29.3	2.5×10^3	1.5×10^4
79	$2,3\text{-}(CH_3)_2C_6H_3OH$	14.5	31.7	1.1×10^3	7.3×10^3
80	$3,4\text{-}(CH_3)_2C_6H_3OH$	13.6	31.2	1.3×10^3	8.4×10^3
81	$3,5\text{-}(CH_3)_2C_6H_3OH$	15.0	32.0	9.6×10^2	6.6×10^3
82	$2,4\text{-}(CH_3)_2C_6H_3OH$	19.5	34.5	3.9×10^2	3.1×10^3
83	$2,6\text{-}(CH_3)_2C_6H_3OH$	15.9	32.5	8.0×10^2	5.7×10^3
85	$4\text{-}(CH_3)_2NC_6H_4OH$	−12.2	18.3	1.3×10^5	4.1×10^5
101	$4\text{-}HO(O)CC_6H_4OH$	30.7	40.9	38	4.6×10^2
102	$4\text{-}ClC_6H_4OH$	29.6	40.2	49	5.6×10^2
106	$4\text{-}CNC_6H_4OH$	29.2	40.0	53	6.0×10^2
110	$2,4,6\text{-}(CH_3)_2C_6H_2OH$	7.8	28.1	3.9×10^3	2.1×10^4
111	$2\text{-}CH_3\text{-}4\text{-}NH_2CH_2\text{-}6\text{-}(CH_3)_3CC_6H_2OH$	9.6	29.1	2.7×10^3	1.6×10^4
112	$2,6\text{-}(CH_3)_2\text{-}4\text{-}NH_2CH_2C_6H_2OH$	7.1	27.8	4.4×10^3	2.3×10^4
113	$2,6\text{-}(CH_3)_2\text{-}4\text{-}COOHCH_2\text{-}C_6H_2OH$	9.0	28.8	3.0×10^3	1.7×10^4
114	$2,6\text{-}(CH_3)_2\text{-}4\text{-}CNC_6H_2OH$	13.0	30.9	1.4×10^3	$.9.2 \times 10^3$
115	$2,3,4,6\text{-}(CH_3)_4C_6HOH$	6.4	27.4	5.0×10^3	2.6×10^4
116	$2,3,5,6\text{-}(CH_3)_4C_6HOH$	10.2	29.4	2.4×10^3	1.4×10^4
117	$(CH_3)_5C_6OH$	−0.5	23.9	1.8×10^4	7.6×10^4
118	$2,3,5,6\text{-}(CH_3)_4\text{-}4\text{-}CH_3OC_6OH$	−0.9	23.7	1.9×10^4	8.0×10^4
126	Indophenol	−12.0	18.4	2.6×10^5	8.0×10^5
130	$\alpha\text{-}(CH_3)_3C_6(OH)(CH_2)_2OC(CH_3)[(CH_2)_3CH(CH_3)]_3CH_3$	−11.3	18.7	1.2×10^5	3.6×10^5
131	$\beta\text{-}(CH_3)_2C_6H(OH)(CH_2)_2OC(CH_3)[(CH_2)_3CH(CH_3)]_3CH_3$	−6.6	20.9	5.3×10^4	1.9×10^5
133	$\delta\text{-}CH_3C_6H_2(OH)(CH_2)_2OC(CH_3)[(CH_2)_3CH(CH_3)]_3CH_3$	−22.0	13.8	6.8×10^5	1.6×10^6

Table 3.9
Rate constants of reaction of phenoxyl radicals with peroxyl radicals in hydrocarbon solutions, $C_{14}H_{11}O_2^{\bullet}$ — peroxyl radical from 9,10-dihydroanthracene

No. of ArOH	Phenoxyl	Peroxyl	T/ K	k/ l mol^{-1} s^{-1}	Ref.
36	$C_6H_5O^{\bullet}$	$C_{14}H_{11}O_2^{\bullet}$	333	1.6×10^8	6
9	$4\text{-}CH_3C_6H_5O^{\bullet}$	$C_{14}H_{11}O_2^{\bullet}$	333	2.4×10^8	6
46	$4\text{-}(CH_3)_3CC_6H_4O^{\bullet}$	$C_{14}H_{11}O_2^{\bullet}$	333	1.1×10^8	6
88	$4\text{-}C_6H_5C_6H_4O^{\bullet}$	$C_{14}H_{11}O_2^{\bullet}$	333	3.9×10^8	6
62	$4\text{-}CH_3OC_6H_4O^{\bullet}$	$C_{14}H_{11}O_2^{\bullet}$	333	7.8×10^8	6
	$3\text{-}CH_3CH_2OC(O)C_6H_4O^{\bullet}$	$C_{14}H_{11}O_2^{\bullet}$	333	3.3×10^8	6
83	$2,6\text{-}(CH_3)_2\text{-}C_6H_3O^{\bullet}$	$C_{14}H_{11}O_2^{\bullet}$	333	1.0×10^8	6
	$3,5\text{-}[(CH_3)_3C]_2\text{-}C_6H_3O^{\bullet}$	$C_{14}H_{11}O_2^{\bullet}$	333	1.5×10^8	6
52	$2,4,6\text{-}[(CH_3)_3C]_3\text{-}C_6H_2O^{\bullet}$	$C_6H_5(CH_3)_2CO_2^{\bullet}$	303	2.0×10^8	7
52	$2,4,6\text{-}[(CH_3)_3C]_3\text{-}C_6H_2O^{\bullet}$	$(CH_3)_2(CN)CO_2^{\bullet}$	333	3.2×10^8	8
62	$2,6\text{-}[(CH_3)_3C]_2\text{-}4\text{-}CH_3OC_6H_2O^{\bullet}$	$(CH_3)_2(CN)CO_2^{\bullet}$	333	7.2×10^8	8
55	$2,6\text{-}[(CH_3)_3C]_2\text{-}4\text{-}(CH_3)_3COC_6H_2O^{\bullet}$	$(CH_3)_2(CN)CO_2^{\bullet}$	333	4.1×10^8	8
70	$2,6\text{-}[(CH_3)_3C]_2\text{-}4\text{-}C_6H_5C_6H_2O^{\bullet}$	$(CH_3)_2(CN)CO_2^{\bullet}$	333	6.0×10^8	8
58	$2,6\text{-}[(CH_3)_3C]_2\text{-}4\text{-}ClC_6H_2O^{\bullet}$	$(CH_3)_2(CN)CO_2^{\bullet}$	333	1.7×10^8	8
	$2,6\text{-}[(CH_3)_3C]_2\text{-}4\text{-}C_6H_5C(O)C_6H_2O^{\bullet}$	$(CH_3)_2(CN)CO_2^{\bullet}$	333	1.9×10^8	8
60	$2,6\text{-}[(CH_3)_3C]_2\text{-}4\text{-}CNC_6H_2O^{\bullet}$	$(CH_3)_2(CN)CO_2^{\bullet}$	333	2.7×10^8	8
	$2,6\text{-}[(CH_3)_3C]_2\text{-}4\text{-}CH_3OC(O)CH_2CH_2C_6H_2O^{\bullet}$	$(CH_3)_2(CN)CO_2^{\bullet}$	333	1.1×10^9	8
29	$1\text{-}C_{10}H_7O^{\bullet}$	$C_6H_5(CH_3)_2CO_2^{\bullet}$	333	6.1×10^8	6

Table 3.10
Rate constants of recombination and disproportionation of phenoxyl radicals in hydrocarbon solutions measured by flash photolysis technique

No. of ArOH	Phenoxyl	T/ K	k/ l mol^{-1} s^{-1} or log $k = A - E/\theta$	Ref.
36	$C_6H_5O^•$	303–345	7.98 + 5.9/ θ	9
7	2-$CH_3C_6H_4O^•$	298	1.9 × 10^8	10
8	3-$CH_3C_6H_4O^•$	285	7.5 × 10^7	11
9	4-$CH_3C_6H_4O^•$	303– 45	9.64 – 0.8/ θ	9
76	2-$CH_3OC_6H_4^•$	293–353	10.9 – 10.4/ θ	12
78	4-$CH_3OC_6H_4O^•$	303–345	8.66 – 6.7/ θ	9
	2-$HOC_6H_4O^•$	298	3.9 × 10^8	10
121	3-$HOC_6H_4O^•$	298	3.5 × 10^9	10
	4-$H_2N(CH_2)_2C_6H_4O^•$	293	1.9 × 10^9	13
	4-$HOOC(NH_2)CHCH_2C_6H_4O^•$	293–298	4.0 × 10^8	14
40	4-$H_2NC_6H_4O^•$	293–298	2.0 × 10^9	15
46	4-$(CH_3)_3CC_6H_4O^•$	303–345	7.12 + 12.1/ θ	9
	2-$C_6H_5C_6H_4O^•$	293–353	10.3 – 9.6/ θ	16
88	4-$C_6H_5C_6H_4O^•$	293–353	11.2 – 11.6/ θ	12
83	2,6-$(CH_3)_2C_6H_3O^•$	303–345	13.16 – 24.7/ θ	9
81	3,5-$(CH_3)_2C_6H_3O^•$	303–345	12.83 – 17.6/ θ	9
	3-HO-5-$CH_3C_6H_3O^•$	298	3.5 × 10^9	10
	2,6-$(CH_3O)_2C_6H_3O^•$	293–353	10.6 – 9.6/θ	12
47	2,4-$[(CH_3)_3C]_2C_6H_3O^•$	293	4.0 × 10^7	17
51	2,6-$[(CH_3)_3C]_2C_6H_3O^•$	293–353	9.0 – 6.2/ θ	16
	3,5-$[(CH_3)_3C]_2C_6H_3O^•$	303–345	9.26 – 5.0/ θ	9
	2,4-$(C_6H_5)_2C_6H_3O^•$	293–253	10.4 – 9.1/ θ	16
75	2-$(CH_3)_3$C-4-$OCH_3C_6H_3O^•$	323	6.8 × 10^4	18
110	2,4,6-$(CH_3)_3C_6H_2O^•$	323	1.0 × 10^8	18
	2,6-$(CH_3)_2$-4-adamantyl-$C_6H_2O^•$	323	9.8 × 10^4	18
	3,4,5-$(CH_3)_3$-$C_6H_2O^•$	303	1.42 × 10^8	19

No. of ArOH	Phenoxyl	*T*/ K	*k*/ l mol^{-1} s^{-1} or log *k* = *A* – *E*/ θ	Ref.
64	2,6-$[(CH_3)_3C]_2$-4-$CH_3C_6H_2O^•$	293	7.4 + 3.6/ θ	20
	2,6-$[(CH_3)_3C]_2$-4-$CH_3OCH_2C_6H_2O^•$	323	3.5	18
	2,6-$[(CH_3)_3C]_2$-4-$CH_3OC(O)(CH_2)_2C_6H_2O^•$	323	7.0×10^2	18
	2,6-$(C_6H_5)_2$-4-$CH_3C_6H_{2O}^•$	293	10.9 – 8.4/ θ	20
95	2,6-$(C_6H_5)_2$-4-$CH_3OC_6H_2O^•$	293	10.4 – 4.4/ θ	20
29	1-$C_{10}H_7O^•$	333	7.0×10^8	6
30	2-$C_{10}H_7O^•$	303	8.0×10^9	9
17	2,2,5,7,8-$(CH_3)_5$-6-O-(*cyclo*-$[OC(CH_2)_2][C_6]$)	323	2.2×10^3	21

Table 3.11
Rate constants of disproportionation of two different phenoxyl radicals in hydrocarbon solutions

No. of ArOH	Phenoxyl	No. of ArOH	Phenoxyl	*T*/ K	*k*/ l mol^{-1} s^{-1}	Ref.
78	4-$CH_3OC_6H_4O^•$	64	2,6-$[(CH_3)_3C]_2$-4-$CH_3C_6H_2O^•$	333	1.0×10^8	22
78	4-$CH_3OC_6H_4O^•$	52	2,4,6-$[(CH_3)_3C]_3$-$C_6H_2O^•$	333	5.3×10^7	22
64	2,6-$[(CH_3)_3C]_2$-4-$CH_3C_6H_2O^•$	52	2,4,6-$[(CH_3)_3C]_3$-C_6H_2O	323	1.0×10^3	23
64	2,6-$[(CH_3)_3C]_2$-4-$CH_3C_6H_2O^•$	58	2,6-$[(CH_3)_3C]_2$-4-$ClC_6H_2O^•$	323	5.3×10^8	23
64	2,6-$[(CH_3)_3C]_2$-4-$CH_3C_6H_2O^•$	53	2,6-$[(CH_3)_3C]_2$-4-$CH_3C(O)C_6H_2O^•$	323	1.6×10^4	23
64	2,6-$[(CH_3)_3C]_2$-4-$CH_3C_6H_2O^•$	75	2-$(CH_3)_3C$-4-$CH_3OC_6H_3$	323	5.6×10^6	23
64	2,6-$[(CH_3)_3C]_2$-4-$CH_3C_6H_2O^•$	17	Chroman, 2,2,5,7,8-pentamethyl-6-oxyl	323	1.9×10^5	23
64	2,6-$[(CH_3)_3C]_2$-4-$CH_3C_6H_2O^•$		Galvinoxyl	323	4.9×10^4	23
64	2,6-$[(CH_3)_3C]_2$-4-$CH_3C_6H_2O^•$		Indoxyl	323	30	23
64	2,6-$[(CH_3)_3C]_2$-4-$CH_3C_6H_2O^•$	24	1,2-Ethane-2-(4′,6′-di-*tert*-butylphenol)-2-(4″,6″- di-*tert*-butylphenoxyl)	323	7.3×10^4	23
64	2,6-$[(CH_3)_3C]_2$-4-$CH_3C_6H_2O^•$	130	α-Tocopheroxyl	323	1.9×10^5	23
64	2,6-$[(CH_3)_3C]_2$-4-$CH_3C_6H_2O^•$		Triethylenglicol-3-(3′-methyl-4′-hydroxy-5′-*tret*-butylphenyl)-3-(3″-methyl-4″-oxyl-5″-*tret*-butylphenyl)dipropionate	323	2.0×10^3	23

No. of ArOH	Phenoxyl	No. of ArOH	Phenoxyl	T/ K	k/ l mol^{-1} s^{-1}	Ref.
64	$2,6\text{-}[(CH_3)_3C]_2\text{-}4\text{-}CH_3C_6H_2O^{\bullet}$		2,6-Dimethyl-4-adamantyl-phenoxyl	323	5.0×10^3	23
53	2,6-Di-*tret*-butyl-4-acetylphenoxyl		$2,6\text{-}[(CH_3)_3C]_2\text{-}4\text{-}CH_3OCH_2C_6H_2O^{\bullet}$	323	2.6×10^2	23
53	2,6-Di-*tret*-butyl-4-acetylphenoxyl		Pentaeritrite-tris-3-(3′,5′-di-*tret*-butyl-4′-hydroxyphenyl)-3-(3″,5″-di-*tret*-butyl-4″-phenoxyl)	323	1.2×10^4	23
53	2,6-Di-*tret*-butyl-4-acetylphenoxyl		Benzene, 1-(3′,5′-di-*tret*-butyl-4′-oxyl)-benzyl-3,5-di-(3′,5′-di-*tret*-butyl-4′-hydroxy)-benzyl	323	7.8×10^2	23
53	2,6-Di-*tret*-butyl-4-acetylphenoxyl		Decane, 1-(3′,5′-di-*tret*-butyl-4′-oxylphenyl)-10-(3″,5″-di-*tret*-butyl-4″-hydroxyphenyl)	323	1.0×10^4	23
17	Chroman, 2,2,5,7,8-pentamethyl-6-oxyl	52	$2,4,6\text{-}[(CH_3)_3C]_3\text{-}C_6H_2O^{\bullet}$	323	5.0×10^2	24
17	Chroman, 2,2,5,7,8-pentamethyl-6-oxyl		2,6-Di-*tret*-butyl-4-(2-acetoxyethylene)phenoxyl	323	6.7×10^4	23

Table 3.12
Activation energies and rate constants of reaction of phenols (Ar_1OH) with O_2 in hydrocarbon solutions calculated by formula 1.17 at $A = 10^{10}$ l mol^{-1} s^{-1} and $E = \Delta H$

No.	Phenol	E/ kJ mol^{-1}	k (333 K)/ l mol^{-1} s^{-1}	k (400 K)/ l mol^{-1} s^{-1}
3	$2\text{-}HOC_6H_4OH$	134.6	2.6×10^{-8}	8.7×10^{-5}
4	$2\text{-}HO\text{-}4\text{-}(CH_3)_3CC_6H_3OH$	139.0	7.1×10^{-9}	3.0×10^{-5}
9	$4\text{-}CH_3C_6H_4OH$	154.4	6.9×10^{-11}	7.4×10^{-7}
10	$(CH_3)_3C_6(OH)(CH_2)_3O$	125.4	4.2×10^{-7}	7.9×10^{-4}
20	$(CH_3)_3C_6(OH)CH_2CH(CH_3)O$	121.0	1.6×10^{-6}	2.3×10^{-3}
25	$1\text{-}HOC_{13}H_9$	132.9	4.4×10^{-8}	1.3×10^{-4}
29	$1\text{-}HOC_{10}H_7$	138.0	9.5×10^{-9}	3.8×10^{-5}

No.	Phenol	E/ kJ mol^{-1}	k (333 K)/ l mol^{-1} s^{-1}	k (400 K)/ l mol^{-1} s^{-1}
30	2-$HOC_{10}H_7$	148.4	4.2×10^{-10}	3.5×10^{-6}
32	1-$HOC_{14}H_9$	149.3	3.2×10^{-10}	2.5×10^{-6}
36	C_6H_5OH	163.6	4.3×10^{-12}	8.1×10^{-8}
38	4-$CH_3COC_6H_4OH$	166.1	2.0×10^{-12}	4.4×10^{-8}
40	4-$NH_2C_6H_4OH$	151.3	1.7×10^{-10}	1.6×10^{-6}
41	4-$C_6H_5CH_2OC_6H_4OH$	144.1	1.5×10^{-9}	8.8×10^{-6}
44	4-$CH_3(CH_2)_3OC_6H_4OH$	142.4	2.5×10^{-9}	1.3×10^{-5}
46	4-$(CH_3)_3CC_6H_4OH$	153.5	9.0×10^{-11}	9.2×10^{-7}
47	2,4-$[(CH_3)_3C]_2C_6H_3OH$	154.1	7.5×10^{-11}	8.0×10^{-7}
74	2-$(CH_3)_3C$-4,6-$(CH_3)_2C_6H_2OH$	150.4	2.3×10^{-10}	1.9×10^{-6}
75	2-$(CH_3)_3C$-4-$CH_3OC_6H_3OH$	138.0	9.5×10^{-9}	3.8×10^{-5}
78	4-$CH_3OC_6H_4OH$	145.6	9.7×10^{-10}	6.2×10^{-6}
79	2,3-$(CH_3)_2C_6H_3OH$	150.1	2.5×10^{-10}	2.1×10^{-6}
80	3,4-$(CH_3)_2C_6H_3OH$	149.2	3.3×10^{-10}	2.6×10^{-6}
81	3,5-$(CH_3)_2C_6H_3OH$	150.6	2.2×10^{-10}	1.8×10^{-6}
82	2,4-$(CH_3)_2C_6H_3OH$	155.1	5.6×10^{-11}	6.3×10^{-7}
83	2,6-$(CH_3)_2C_6H_3OH$	151.5	1.6×10^{-10}	1.5×10^{-6}
85	4-$(CH_3)_2NC_6H_4OH$	123.4	7.7×10^{-7}	1.3×10^{-3}
110	2,4,6-$(CH_3)_3C_6H_2OH$	143.4	1.9×10^{-9}	1.0×10^{-5}
130	α-$(CH_3)_3C_6(OH)(CH_2)_2OC(CH_3)[(CH_2)_3CH(CH_3)]_3CH_3$	124.3	5.9×10^{-7}	1.0×10^{-3}
131	β-$(CH_3)_2C_6H(OH)(CH_2)_2OC(CH_3)[(CH_2)_3CH(CH_3)]_3CH_3$	129.0	1.4×10^{-7}	3.3×10^{-4}
133	δ-$CH_3C_6H_2(OH)(CH_2)_2OC(CH_3)[(CH_2)_3CH(CH_3)]_3CH_3$	113.6	1.5×10^{-5}	1.4×10^{-2}

Table 3.13
Activation energies and rate constants of reaction of sterically hindered phenols (Ar_2OH) with O_2 in hydrocarbon solutions calculated by formula 1.17 at $A = 10^{10}$ l mol^{-1} s^{-1} and $E = \Delta H$

No.	Phenol	E/ kJ mol^{-1}	k (400 K)/ l mol^{-1} s^{-1}	k (500 K)/ l mol^{-1} s^{-1}
1	1,3,5-[3′,5′-[$(CH_3)_2C]_2$-4′-HO-$C_6H_5CH_2]_3C_6H_3$	141.5	9.9×10^{-9}	5.1×10^{-5}
24	2,2′-CH_2-[4,6-[$(CH_3)_3C]_2C_6H_2OH]_2$	137.5	2.2×10^{-8}	8.6×10^{-5}
31	C-[4-$(CH_2)_2COOCH_2$-2,6-[$(CH_3)_3C]_2C_6H_2OH]_4$	137.6	4.4×10^{-8}	1.7×10^{-6}
49	2,4-[$(CH_3)_3C]_2$-6-CH_3-C_6H_2OH	150.5	2.2×10^{-10}	1.9×10^{-6}
50	4,4′-[2,6-[$(CH_3)_3C]_2C_6H_2OH]_2$	139.2	1.3×10^{-8}	5.8×10^{-5}
51	2,6-[$(CH_3)_3C]_2C_6H_3OH$	143.3	1.9×10^{-9}	1.1×10^{-5}
52	2,6-[$(CH_3)_3C]_2$-4-$CH_3CONHC_6H_2OH$	123.3	7.9×10^{-7}	1.3×10^{-3}
53	2,6-[$(CH_3)_3C]_2$-4-$CH_3COC_6H_2OH$	142.1	2.8×10^{-9}	1.4×10^{-5}
54	2,6-[$(CH_3)_3C]_2$-4-$C_6H_5CH_2C_6H_2OH$	136.3	1.6×10^{-8}	5.8×10^{-5}
55	2,6-[$(CH_3)_3C]_2$-4-$(CH_3)_3COC_6H_2OH$	127.9	2.0×10^{-7}	4.4×10^{-4}
56	2,6-[$(CH_3)_3C]_2$-4-$(CH_3)_3COC(O)C_6H_2OH$	144.3	1.4×10^{-9}	8.4×10^{-6}
57	2,6-[$(CH_3)_3C]_2$-4-$HOOCC_6H_2OH$	145.4	1.0×10^{-9}	6.5×10^{-6}
58	2,6-[$(CH_3)_3C]_2$-4-ClC_6H_2OH	141.1	3.8×10^{-9}	1.8×10^{-5}
60	2,6-[$(CH_3)_3C]_2$-4-NCC_6H_2OH	149.0	3.5×10^{-10}	2.7×10^{-6}
61	2,6-[$(CH_3)_3C]_2$-4-$H(O)CC_6H_2OH$	144.4	1.4×10^{-9}	8.2×10^{-6}
62	2,6-[$(CH_3)_3C]_2$-4-$CH_3OC_6H_2OH$	127.3	2.4×10^{-7}	5.0×10^{-4}
68	2,6-[$(CH_3)_3C]_2$-4-$NO_2C_6H_2OH$	154.6	6.5×10^{-11}	7.1×10^{-7}
70	2,6-[$(CH_3)_3C]_2$-4-$C_6H_5C_6H_2OH$	134.3	2.9×10^{-8}	9.3×10^{-5}
73	2,6-[$(CH_3)_3C]_2$-4-$C_6H_5SC_6H_2OH$	143.0	2.1×10^{-9}	1.1×10^{-5}
89	2,6-$(C_6H_5)_2$-4-$NH_2CH_2C_6H_2OH$	126.5	3.0×10^{-7}	6.1×10^{-4}
91	2,6-$(C_6H_5)_2$-4-$HOOCCH_2C_6H_2OH$	122.5	1.0×10^{-6}	1.6×10^{-3}
94	2,6-$(C_6H_5)_2$-4-$CH_3SC_6H_2OH$	137.0	1.3×10^{-8}	4.9×10^{-5}
95	2,6-$(C_6H_5)_2$-4-$CH_3OC_6H_2OH$	124.6	5.4×10^{-6}	9.6×10^{-4}
122	Si[2,6-[$(CH_3)_3C]_2$-4-$CH_2CH_2OC_6H_2OH]_4$	137.5	4.4×10^{-8}	1.7×10^{-4}
123	4,4-CH_2SCH_2-[2,6-[$(CH_3)_3C]_2C_6H_2OH]_2$	136.5	3.0×10^{-8}	5.5×10^{-5}

Table 3.14
Rate constants of reaction of phenols with hydroperoxides

No.	Phenol	Solvent	*T*/ K	*k*/ l mol^{-1} s^{-1}	*E*/ kJ mol^{-1}	log (*A*/ l mol^{-1} s^{-1})	Ref.
4	2-HOC_6H_4OH	C_6H_5Cl	413	1.8×10^{-2}	102.4	10.24	25
9	4-$CH_3C_6H_4OH$	C_6H_5Cl	413	2.0×10^{-4}	109.5	10.15	26
26	4-HOC_6H_4OH	C_6H_5Cl	413	2.0×10^{-3}	110.3	11.27	25
29	1-$C_{10}H_7OH$	C_6H_5Cl	413	6.7×10^{-4}	101.2	9.63	26
52	2,4,6-$[(CH_3)_3C]_3C_6H_2OH$	$C_{10}H_{22}$	413	7.8×10^{-5}	108.3	9.60	26
64	2,6-$[(CH_3)_3C]_2$-4-CH_3-C_6H_2OH	$C_6H_5C_2H_5$	393	3.0×10^{-4}	—	—	27
64	2,6-$[(CH_3)_3C]_2$-4-CH_3-C_6H_2OH	C_6H_6	418	4.1×10^{-3}	—	—	28
78	4-$CH_3OC_6H_4OH$	C_6H_5Cl	413	4.3×10^{-3}	97.4	9.95	26
	2-CH_3-4,6-$[(CH_3)_3C]_2$-C_6H_2OH	C_6H_6	418	7.0×10^{-3}	—	—	28
	2,6-$(CH_3O)_2C_6H_3OH$	C_6H_5Cl	413	6.5×10^{-3}	95.7	9.91	26
	2,3-$C_{10}H_6(OH)_2$	C_6H_5Cl	413	4.5×10^{-3}	107.8	11.33	26
	2,2'-CH_2-$[2,6[(CH_3)_3C]_2C_6H_2OH]_2$	C_6H_5Cl	413	4.2×10^{-4}	108.7	10.36	26

Table 3.15
Rate constants, preexponential factors and activation energies of decay of quinolide peroxides $ROOR_i$ in benzene at 363 K[29]

$ROOR_i$	k/ s^{-1}	log (A/ s^{-1})	E/ kJ mol^{-1}
$(CH_3)_3COOR_1$	6.6×10^{-5}	14.0	127.5
$CH_3(CH_2)_2(CH_3)_2COOR_1$	6.3×10^{-5}	14.1	127.9
$C_6H_5(CH_3)_2COOR_1$	1.9×10^{-4}	15.5	134.6
R_1OOR_2	1.4×10^{-4}	15.4	134.6
$(CH_3)_3COOR_2$	1.8×10^{-6}	14.0	137.9
$CH_3(CH_2)_2(CH_3)_2COOR_2$	1.9×10^{-6}	15.5	148.0
$C_6H_5(CH_3)_2COOR_2$	6.8×10^{-6}	15.7	146.3
R_2OOR_2	1.7×10^{-5}	14.5	134.6
$(CH_3)_3COOR_3$	6.6×10^{-6}	13.6	137.9
$(CH_3)_3COOR_4$	2.5×10^{-5}	12.8	121.2
$(CH_3)_3COOR_5$	3.7×10^{-5}	14.4	131.7

R_1 = *cyclo*-[CH=C(*tert*-$C(CH_3)_3$)CH=C(*tert*-$C(CH_3)_3$)C(O)C(*tert*-$C(CH_3)_3$)-]

R_2 = *cyclo*-[CH=C(*tert*-$C(CH_3)_3$)C(O)C(*tert*-$C(CH_3)_3$)=CHC(*tert*-$C(CH_3)_3$)-]

R_3 = *cyclo*-[CH=C(*tert*-$C(CH_3)_3$)C(O)C(*tert*-$C(CH_3)_3$)=CHC(CH_3)-]

R_4 = *cyclo*-[CH=C(*tert*-$C(CH_3)_3$)C(O)C(*tert*-$C(CH_3)_3$)=CHC(*tert*-$OC(CH_3)_3$))-]

R_5 = *cyclo*-[CH=C(*tert*-$C(CH_3)_3$)C(O)C(*tert*-$C(CH_3)_3$)=CHC(C_6H_5)-]

Table 3.16
The values of stoichiometric coefficients of chain termination by phenols in oxidizing substances RH

No.	Phenol	RH	*T*/ K	*f*	Ref.
3	$2\text{-}HOC_6H_4OH$	$C_6H_5CH(CH_3)_2$	338	2.3	30
4	$2\text{-}HO\text{-}4\text{-}(CH_3)_3C\text{-}C_6H_3OH$	$C_6H_5CH(CH_3)_2$	338	2.2	30
4	$2\text{-}HO\text{-}4\text{-}(CH_3)_3C\text{-}C_6H_3OH$	$C_6H_5CH(CH_3)_2$	338	2.0	31
5	$2\text{-}HO\text{-}3,5\text{-}[(CH_3)_3C]_2C_6H_2OH$	$C_6H_5CH(CH_3)_2$	338	2.0	30
7	$2\text{-}CH_3C_6H_4OH$	$C_6H_5CH(CH_3)_2$	338	2.1	30
7	$2\text{-}CH_3C_6H_4OH$	$C_6H_5CH(CH_3)_2$	338	2.2	31
9	$4\text{-}CH_3C_6H_4OH$	$C_6H_5CH(CH_3)_2$	338	2..5	30
9	$4\text{-}CH_3C_6H_4OH$	$C_6H_5CH(CH_3)_2$	338	2.2	31
25	$1\text{-}HOC_{13}H_9$	9,10-Dihydroanthracene	333	3.6	32
26	$4\text{-}HOC_6H_4OH$	$C_6H_5CH(CH_3)_2$	338	1.8	30
26	$4\text{-}HOC_6H_4OH$	9,10-Dihydroanthracene	333	1.75	32
28	$5\text{-}HOC_{10}H_6OH$	9,10-Dihydroanthracene	333	2.05	32
29	$1\text{-}HOC_{10}H_7$	$C_6H_5CH(CH_3)_2$	338	2.0	30
29	$1\text{-}HOC_{10}H_7$	9,10-Dihydroanthracene	333	2.08	32
29	$1\text{-}HOC_{10}H_7$	*cyclo*-$C_6H_{10}O$	348	2.0	33
29	$1\text{-}HOC_{10}H_7$	*cyclo*-$C_6H_{11}OH$	348	2.7	34
29	$1\text{-}HOC_{10}H_7$	$C_6H_5CH(CH_3)_2$	333	2.1	35
30	$2\text{-}HOC_{10}H_7$	$C_6H_5CH(CH_3)_2$	338	1.8	30
30	$2\text{-}HOC_{10}H_7$	$C_6H_5CH(CH_3)_2$	338	2.1	31
36	C_6H_5OH	$C_6H_5CH(CH_3)_2$	338	2.1	30
36	C_6H_5OH	$C_6H_5CH(CH_3)_2$	338	2.0	31
45	$2\text{-}(CH_3)_3CC_6H_4OH$	$C_6H_5CH(CH_3)_2$	338	2.1	30
46	$4\text{-}(CH_3)_3CC_6H_4OH$	$C_6H_5CH(CH_3)_2$	338	2.3	30
47	$2,4\text{-}[(CH_3)_3C]_2C_6H_3OH$	$C_6H_5CH(CH_3)_2$	338	2.3	30
49	$2,4\text{-}[(CH_3)_3C]_2\text{-}6\text{-}CH_3\text{-}C_6H_2OH$	$C_6H_5CH(CH_3)_2$	338	2.0	30
49	$2,4\text{-}[(CH_3)_3C]_2\text{-}6\text{-}CH_3\text{-}C_6H_2OH$	$C_6H_5CH(CH_3)_2$	338	2.0	31
49	$2,4\text{-}[(CH_3)_3C]_2\text{-}6\text{-}CH_3\text{-}C_6H_2OH$	$C_6H_5CH(CH_3)_2$	333	2.1	35
51	$2,6\text{-}[(CH_3)_3C]_2C_6H_3OH$	$C_6H_5CH(CH_3)_2$	338	2.0	30
52	$2,4,6\text{-}[(CH_3)_3C]_3C_6H_2OH$	$C_6H_5CH(CH_3)_2$	338	1.8	30

No.	Phenol	RH	*T*/ K	*f*	Ref.
52	2,4,6-$[(CH_3)_3C]_3C_6H_2OH$	$C_6H_5CH(CH_3)_2$	333	2.1	35
75	2-$(CH_3)_3C$-4-$CH_3OC_6H_3OH$	$C_6H_5CH(CH_3)_2$	338	2.1	30
78	4-$CH_3OC_6H_4OH$	$C_6H_5CH(CH_3)_2$	338	2.0	30
78	4-$CH_3OC_6H_4OH$	$C_6H_5CH(CH_3)_2$	338	2.1	31
82	2,4-$(CH_3)_2C_6H_3OH$	$C_6H_5CH(CH_3)_2$	338	2.6	30
82	2,4-$(CH_3)_2C_6H_3OH$	$C_6H_5CH(CH_3)_2$	338	2.0	31
83	2,6-$(CH_3)_2C_6H_3OH$	$C_6H_5CH(CH_3)_2$	338	2.0	30
102	4-ClC_6H_4OH	$C_6H_5CH(CH_3)_2$	338	1.7	30
102	4-ClC_6H_4OH	$C_6H_5CH(CH_3)_2$	338	1.0	31
108	2-CH_3-4-$(CH_3)_3C$-C_6H_3OH	$C_6H_5CH(CH_3)_2$	338	1.9	30
110	2,4,6-$(CH_3)_3C_6H_2OH$	$C_6H_5CH(CH_3)_2$	338	2.2	30
119	[3,5-$[(CH_3)_3C]_2$-4-$HOC_6H_2(CH_2)_2COOCH_2]_4C$	$C_6H_5CH(CH_3)_2$	333	2.1	35
123	4,4′-CH_2SCH_2-[2,6-$[(CH_3)_3C]_2C_6H_2OH]_2$	$C_6H_5CH(CH_3)_2$	333	4.8	35
	2-CH_3-4-ClC_6H_3OH	$C_6H_5CH(CH_3)_2$	338	1.7	30
	2,6-$(CH_3)_2$-4-ClC_6H_2OH	$C_6H_5CH(CH_3)_2$	338	1.3	30
	2,6-$(CH_3)_2$-4-CNC_6H_2OH	$C_6H_5CH(CH_3)_2$	338	1.7	30
	2-CH_3-4-$(CH_3)_3CC_6H_3OH$	$C_6H_5CH(CH_3)_2$	338	2.4	30
	2-CH_3-6-$(CH_3)_3CC_6H_3OH$	$C_6H_5CH(CH_3)_2$	338	1.9	30
	2-CH_3O-4-$CH_3C_6H_3OH$	$C_6H_5CH(CH_3)_2$	338	1.3	30
	2,6-$(CH_3O)_2C_6H_3OH$	$C_6H_5CH(CH_3)_2$	338	1.4	30
	2-CH_3-4,6-$[(CH_3)_3C]_2C_6H_2OH$	$C_6H_5CH(CH_3)_2$	338	1.5	30
	2-$(CH_3)_3C$-4-HOC_6H_3OH	$C_6H_5CH(CH_3)_2$	338	1.1	30
	2,5-$[(CH_3)_3C]_2$-4-HOC_6H_2OH	$C_6H_5CH(CH_3)_2$	338	0.85	31
	2,3-$(OH)_2C_6H_3OH$	$C_6H_5CH(CH_3)_2$	338	2.5	30
	2-HO-4-$CH_3CH_2CH_2C(O)OC_6H_3OH$	$C_6H_5CH(CH_3)_2$	338	2.0	30
	Pyren, 3,10-dihydroxy	9,10-Dihydroanthracene	333	1.9	32
	Pyren, 3,8-dihydroxy	9,10-Dihydroanthracene	333	1.94	32
	2,2′-[4,6-$[(CH_3)_3C]_2C_6H_2OH]_2$	$C_6H_5CH(CH_3)_2$	338	1.4	30
	2,2′-CH_2-[4-$CH_3C_6H_3OH]_2$	$C_6H_5CH(CH_3)_2$	333	4.0	35
	4,4′-$C(CH_3)_2$-$[C_6H_4OH]_2$	$C_6H_5CH(CH_3)_2$	338	4.1	31
	4,4′-N_2-$[C_6H_4OH]_2$	$C_6H_5CH(CH_3)_2$	338	2.0	31

REFERENCES

1. **Denisov, E. T.**, The estimation of O—H-bonds dissociation energies of phenols on kinetics data in the scope of parabolic model, *Zh. Fiz. Khim.*, (in press).
2. **Suryan, M. M., Kafafi, S. A., Stein, S. E.**, Dissociation of substituted anisoles: Substituent effects on bond strengths, *J. Am. Chem. Soc.*, 111, 4594, 1989.
3. **Bordwell, F. G., Cheng, J. P.**, Substituent effects on the stabilities of phenoxyl radicals and the acidities of phenoxyl radical cations, *J. Am. Chem. Soc.*, 113, 1736, 1991.
4. **Mahoney, L. R., Ferris, F. C., DaRooge, M. A.**, Calorimetric study of the 2,4,6-tri-*tert*-butyl phenoxy radical in solution, *J. Am. Chem. Soc.*, 91, 3883, 1969.
5. **Buchachenko, A. L., Vasserman, A. M.**, *Stable radicals*, Khimiya, Moscow, 1973, 366 (in Russian).
6. **Mahoney, L. R., DaRooge, M. A.** The kinetic behaviour and thermochemical properties of phenoxy radicals. *J. Am. Chem. Soc.*, 97, 4722, 1975.
7. **Griva, A. P., Denisov, E. T.**, Kinetics of the reactions of 2,4,6-tri-*tert*-butylphenoxyl with cumene hydroperoxide, cumylperoxyl radicals, and molecular oxygen, *Int. J. Chem. Kinet.*, 5, 869, 1973.
8. **Rubtsov, V. I., Roginskii, V. A., Dubinskii, V. Z., Miller, V. B.**, Rate constants of phenoxy radical reactions in hydrocarbon oxidation inhibited with 2,4,6-tri-*tert*-butylphenol, *Kinet. Katal.* , 19, 1140, 1978.
9. **Mahoney, L. R., Weiner, S. A.**, A mechanistic study of the dimerisation of phenoxyl radicals, *J. Am. Chem. Soc.*, 94, 585, 1972.
10. **Khudyakov, I. V., Kuzmin, V. A., Emanuel, N. M.**, Decay kinetics of aryloxy and semiquinone radicals in the presence of copper ions, *Int. J. Chem. Kinet.*, 10, 1005, 1978.
11. **Khudyakov, I. V., Kuzmin, V. A.**, The absolute values of rate constants of decay of phenoxyl radicals in water solutions, *Khim. Vysok. Energii*, 7, 331, 1973.
12. **Levin, P. P., Khudyakov, I. V., Kuzmin, V. A.**, Effect of solvent viscosity on the kinetics of reversible dimerisation of the phenoxy radicals, *Int. J. Chem. Kinet.*, 12, 147, 1980.
13. **Feitelson, J., Hayon, E.**, Electron ejection and electron capture by phenolic compounds., *J. Phys. Chem.*, 77, 10, 1973.
14. **Solar, S., Solar, W., Getoff, N.**, Reactivity of OH with tyrosini in aqueous solution studied by pulse radiolysis, *J. Phys. Chem.*, 88, 2091, 1984.
15. **Tripathi, G. N. R., Schuter, R. H.**, Resonance Raman studies of pulse radiolytically produced *p*-aminophenoxyl radicals, *J. Phys. Chem.*, 88, 1706, 1984.
16. **Levin, P. P., Khudyakov, I. V., Kuzmin, V. A.**, The study of decay of mono- and bisubstituted phenoxyl radicals by flash photolysis technique, *Izv. Akad. Nauk SSSR, Ser. Khim.*, N2, 255, 1980.
17. **Land, E. J., Porter, G.**, Primary photochemical processes in aromatic molecules. Part 7. Spectra and kinetics of some phenoxyl derivatives, *Trans. Faraday Soc.*, 59, 2016, 1963.
18. **Krashennikova, G. A., Roginskii, V. A.**, ESR-spectra and kinetics of disproportionation of substituted phenoxyl radicals. 4. Rate constants of crossdisproportionation, *Kinet. Katal.*, 30, 606, 1989.
19. **Weiner, S. A.**, A steady state technique for measuring phenoxy radical termination constants, *J. Am. Chem. Soc.*, 94, 581, 1972.
20. **Khudyakov, I. V., Yasmenko, A. I., Kuzmin, V. A., Levin, P. P., Khardin, A. P.**, The study of free radicals of inhibitors by pulse photolysis technique, *Neftekhimiya*, 18, 716, 1978.
21. **Roginskii, V. A., Krasheninnikova, G. A.**, Thesis of 2 Vsesoyuznaya conference "*Bioantioxidant*", Chernogolovka, OICP, 1986, 1, 20 (in Russian).
22. **Mahoney, L. R., DaRooge, M. A.**, Inhibition of free-radical reactions. The synergistic effect of 2,6-di-*tert*-butylphenols on hydrocarbon oxidation retarded by 4-methoxyphenol. *J. Am. Chem. Soc.*, 89, 5619, 1967.
23. **Krasheninnikova, G. A., Roginskii, V. A.**, ESR-spectra and kinetics of disproportionation of substituted phenoxy radicals. IV. Rate constants of crossdisproportionation, *Kinet. Katal.*, 30, 606, 1989.
24. **Roginskii, V. A., Krasheninnikova, G. A.**, ESR-spectra and kinetics of disproportionation of substituted phenoxy radicals. III. Crossdisproportionation in the system with two phenols, *Kinet. Katal.*, 28, 305, 1987.
25. **Nikolaevskii, A. N., Fillipenko, T. A., Peicheva, A. I., Kucher, R. V.**, The reaction of cumylhydroperoxide with diatomic nonsubstituted phenols, *Neftekhimiya*, 16, 758, 1976.
26. **Martemyanov, V. S., Denisov, E. T., Samoylova, L. A.**, Reaction of phenols with cumyl hydroperoxide, *Izv. Akad. Nauk SSSR, Ser. Khim.*, 1972, 1039.
27. **Heberger, K.**, On the interaction of hydroperoxide and inhibitor molecule, *Int. J. Chem. Kinet.*, 17, 271, 1985.
28. **Zikmund, Z., Bradilova, J., Pospishil J.**, Antioxidants and stabilizers. 42. Reaction of tert-butyl hydroperoxide with mononuclear and binuclear phenols. *J. Polym. Sci., Polymer Symp.*, N40, 271, 1973.
29. **Roginsky, V. A., Dubinsky, V. Z., Shlyapnikova, I. A., Miller, V. B.**, Effectiveness of phenol antioxidants and the properties of quinolide peroxides, *Eur. Polym. J.*, 13, 1043, 1977.
30. **Horswill, E.C., Howard, J.A., Ingold, K.U.**, The stoichiometry of oxidation of substituted phenols by peroxy radicals, *Can. J. Chem.*,44, 985, 1966.
31. **Boozer, C.E., Hammond, G.S., Hamilton, C.E., Sen, J.N.**, Air oxidation of hydrocarbons. 2. The stoichiometry and fate of inhibitors in benzene and chlorobenzene, *J. Am. Chem. Soc.*, 77, 3233, 1955.

32. **Mahoney, L. R.**, Inhibition of free radical reactions. 2. Kinetic study of the reaction of peroxy radicals with hydroquinones and hindered phenols. *J. Am Chem. Soc.*, 88, 3035, 1966.
33. **Alexandrov, A.L., Denisov, E.T.**, Rate constants of reaction of peroxyl radicals in oxidizing cyclohexanone, *Kinet. Katal.*, 10, 904, 1969.
34. **Alexandrov,A.L., Sapacheva, T.I., Shuvalov, V.F.** Rate constants of initiation by AIBN and the stoichiometric factor for 1-naphtol in cyclohexanole, *Izv. Akad. Nauk SSSR, Ser. Khim.*, 1969, 955.
35. **Tsepalov, V. F.**, Study of elementary reactions of liquid–phase oxidation of alkylaromatic hydrocarbons, Doct. Sci. (Chem.) Thesis Dissertation, Inst. Chem. Phys., Chernogolovka, 1975, 31 (in Russian).

Chapter 4

BOND DISSOCIATION ENERGIES AND RATE CONSTANTS OF REACTIONS OF AROMATIC AMINES

Table 4.1
Dissociation energies of N—H-bonds in aromatic amines

No.	Name and formula of amine	Mol. wt.	$D_{N—H}$/ kJ mol^{-1}	Ref.
134	Carbasol; $1,2\text{-}C_6H_4NH\text{-}1',2'\text{-}C_6H_4$	167.22	371.6	1
135	Di-2-naphthyamine; $(2\text{-}C_{10}H_7)_2NH$	269.36	360.2	1
136	Diphenylamine; $(C_6H_5)_2NH$	169.23	364.7	1
137	Diphenylamine, 4,4′-dibrom; $(4\text{-}BrC_6H_4)_2NH$	327.02	364.2	1
138	Diphenylamine, 4,4′-dimethoxy; $(4\text{-}CH_3OC_6H_4)_2NH$	229.28	348.6	1
139	Diphenylamine, 4,4′-dimethyl; $(4\text{-}CH_3C_6H_4)_2NH$	197.28	357.5	1
140	Diphenylamine, 4,4′-di-*tert*-butyl; $(4\text{-}(CH_3)_3CC_6H_4)_2NH$	281.45	358.8	1
141	Diphenylamine, 4-methoxy; $4\text{-}CH_3OC_6H_4NHC_6H_5$	199.26	355.9	1
142	Diphenylamine, 4-nitro; $4\text{-}NO_2C_6H_4NHC_6H_4$	214.23	372.9	1
143	Diphenylamine, 4-*tert*-butyl; $4\text{-}(CH_3)_3CC_6H_4NHC_6H_5$	225.33	360.3	1

No.	Name and formula of amine	Mol. wt.	D_{N-H}/ kJ mol^{-1}	Ref.
144	1-Naphthylamine; $1\text{-}C_{10}H_7NH_2$	156.19	374.7	2
145	2-Naphthylamine; $2\text{-}C_{10}H_7NH_2$	156.19	379.5	2
146	*p*-Phenylendiamine, N,N′-di-2-naphthyl; $4\text{-}(2\text{-}C_{10}H_7NH)C_6H_4NH\text{-}2'\text{-}C_{10}H_7$	360.47	346.6	2
147	*p*-Phenylendiamine, N,N′-dioctyl; $4\text{-}C_8H_{17}NHC_6H_4NHC_8H_{17}$	332.57	346.9	2
148	*p*-Phenylendiamine, N,N′-diphenyl; $4\text{-}C_6H_5NH\text{-}4\text{-}C_6H_4NHC_6H_5$	260.35	355.9	2
149	*p*-Phenylendiamine, N,N′-di(4-isopropyl)phenyl; $4\text{-}(CH_3)_2CHC_6H_4NH\text{-}4\text{-}C_6H_4NHC_6H_4CH(CH_3)_3$	344.51	333.6	2
150	*p*-Phenylendiamine, N-phenyl,N′-isopropyl; $4\text{-}C_6H_5NHC_6H_4NHCH(CH_3)_2$	226.33	349.2	2
151	Phenylamine, N-1-(3,7-di-*tert*-butyl)naphthyl; $1\text{-}(3,7\text{-}[(CH_3)_3C]_2C_{10}H_5)NHC_6H_5$	325.50	344.9	2
152	Phenyl-1-naphthylamine; $1\text{-}C_{10}H_7NHC_6H_5$	219.30	357.1	2
153	Phenyl-2-naphthylamine; $2\text{-}C_{10}H_7NHC_6H_5$	219.30	362..9	1
154	4-Phenyloxyphenyl-2-naphthylamine; $2\text{-}C_{10}H_7NH(4'\text{-}C_6H_5O)C_6H_4$	311.40	349.5	2
155	4-(Spirotetrahydrofuran-2′)spirocyclohexyl-1,2,3,4-tetrahydroquinoline; 1,2-*cyclo*-[C(*cyclo*-$CH_2CH_2O)CH_2C(NH\text{-}($*cyclo*-$C_6H_{11}))]C_6H_4$	256.37	369.8	2
156	4-*tert*-Butylphenyl-1-(4′-*tert*-butyl)naphthylamine; $1\text{-}[NHC_6H_4\text{-}4\text{-}C(CH_3)_3]\text{-}4'\text{-}(CH_3)_3CC_{10}H_6$	331.51	352.1	2
157	1,2,3,4-Tetrahydroquinoline, 2,2,4-trimethyl; 1,2-*cyclo*-$[C(CH_3)CH_2C(CH_3)_2NH]C_6H_4$	174.27	368.5	2

Table 4.2
Enthalpies, activation energies and rate constants of reaction of peroxyl radicals ($RO_2^•$) with aromatic amines calculated by formulas 1.15–1.17 and 1.20. The values of A, br_e and α, see Table 1.6

No. of AmH	Amine	$RO_2^•$	ΔH/ kJ mol^{-1}	E/ kJ mol^{-1}	k (333 K)/ l mol^{-1} s^{-1}	k (400 K)/ l mol^{-1} s^{-1}
135	$(2\text{-}C_{10}H_7)_2NH$	$HO_2^•$	−8.8	15.7	3.4×10^5	8.9×10^5
135	$(2\text{-}C_{10}H_7)_2NH$	*sec*-$RO_2^•$	−5.3	17.3	1.9×10^5	5.5×10^5
135	$(2\text{-}C_{10}H_7)_2NH$	*tert*-$RO_2^•$	1.6	20.6	5.9×10^4	2.0×10^5
136	$(C_6H_5)_2NH$	$HO_2^•$	−4.3	17.8	1.6×10^5	4.7×10^5
136	$(C_6H_5)_2NH$	*sec*-$RO_2^•$	−0.8	19.5	8.7×10^4	2.8×10^5
136	$(C_6H_5)_2NH$	*tert*-$RO_2^•$	6.1	22.8	2.6×10^4	1.1×10^5
137	$(4\text{-}BrC_6H_4)_2NH$	$HO_2^•$	−4.8	17.6	1.7×10^5	5.0×10^5
137	$(4\text{-}BrC_6H_4)_2NH$	*sec*-$RO_2^•$	−1.3	19.2	9.7×10^4	3.1×10^5
137	$(4\text{-}BrC_6H_4)_2NH$	*tert*-$RO_2^•$	−5.6	17.2	2.0×10^5	5.7×10^5
138	$(4\text{-}CH_3OC_6H_4)_2NH$	$HO_2^•$	−0.4	10.7	2.1×10^6	4.0×10^6
138	$(4\text{-}CH_3OC_6H_4)_2NH$	*sec*-$RO_2^•$	−16.9	12.1	1.3×10^6	2.6×10^6
138	$(4\text{-}CH_3OC_6H_4)_2NH$	*tert*-$RO_2^•$	−10.0	15.2	4.1×10^5	1.0×10^6
139	$(4\text{-}CH_3C_6H_4)_2NH$	$HO_2^•$	−11.5	14.5	5.3×10^5	1.3×10^6
139	$(4\text{-}CH_3C_6H_4)_2NH$	*sec*-$RO_2^•$	−8.0	16.1	3.0×10^5	7.9×10^5
139	$(4\text{-}CH_3C_6H_4)_2NH$	*tert*-$RO_2^•$	−1.1	19.3	9.4×10^4	3.0×10^5
140	$(4\text{-}(CH_3)_3CC_6H_4)_2NH$	$HO_2^•$	−10.2	15.1	4.3×10^5	1.1×10^6
140	$(4\text{-}(CH_3)_3CC_6H_4)_2NH$	*sec*-$RO_2^•$	−6.7	16.7	2.4×10^5	6.6×10^5
140	$(4\text{-}(CH_3)_3CC_6H_4)_2NH$	*tert*-$RO_2^•$	0.2	19.9	7.6×10^4	2.5×10^5
141	$4\text{-}CH_3OC_6H_4NHC_6H_5$	$HO_2^•$	−13.1	13.8	6.8×10^5	1.6×10^6
141	$4\text{-}CH_3OC_6H_4NHC_6H_5$	*sec*-$RO_2^•$	−9.6	15.4	3.8×10^5	9.7×10^5
141	$4\text{-}CH_3OC_6H_4NHC_6H_5$	*tert*-$RO_2^•$	−2.7	18.6	1.2×10^5	3.7×10^5
142	$4\text{-}NO_2C_6H_4NHC_6H_4$	$HO_2^•$	3.9	21.8	3.8×10^4	1.4×10^5
142	$4\text{-}NO_2C_6H_4NHC_6H_4$	*sec*-$RO_2^•$	7.4	23.5	2.1×10^4	8.5×10^4
142	$4\text{-}NO_2C_6H_4NHC_6H_4$	*tert*-$RO_2^•$	14.3	27.0	5.8×10^3	3.0×10^4
143	$4\text{-}(CH_3)_3CC_6H_4NHC_6H_5$	$HO_2^•$	− 8.7	15.8	3.3×10^5	8.6×10^5

No. of AmH	Amine	$RO_2^•$	$ΔH$/ kJ mol^{-1}	E/ kJ mol^{-1}	k (333 K)/ l mol^{-1} s^{-1}	k (400 K)/ l mol^{-1} s^{-1}
143	$4-(CH_3)_3CC_6H_4NHC_6H_5$	*sec*-$RO_2^•$	−5.2	17.4	1.9×10^5	5.3×10^5
143	$4-(CH_3)_3CC_6H_4NHC_6H_5$	*tert*-$RO_2^•$	1.7	20.7	5.7×10^4	2.0×10^5
144	$1-C_{10}H_7NH_2$	$HO_2^•$	5.7	22.6	5.7×10^4	2.2×10^5
144	$1-C_{10}H_7NH_2$	*sec*-$RO_2^•$	9.2	24.4	3.0×10^4	1.3×10^5
144	$1-C_{10}H_7NH_2$	*tert*-$RO_2^•$	16.1	28.0	8.1×10^3	4.4×10^4
145	$2-C_{10}H_7NH_2$	$HO_2^•$	10.5	25.1	2.3×10^4	1.1×10^5
145	$2-C_{10}H_7NH_2$	*sec*-$RO_2^•$	14.0	26.9	1.2×10^4	6.1×10^4
145	$2-C_{10}H_7NH_2$	*tert*-$RO_2^•$	20.9	30.6	3.2×10^3	2.0×10^4
146	$2-C_{10}H_7NHC_6H_4NH-2'-C_{10}H_7$	$HO_2^•$	−22.4	9.8	5.8×10^6	1.1×10^7
146	$2-C_{10}H_7NHC_6H_4NH-2'-C_{10}H_7$	*sec*-$RO_2^•$	−18.9	11.3	3.4×10^6	6.7×10^6
146	$2-C_{10}H_7NHC_6H_4NH-2'-C_{10}H_7$	*tert*-$RO_2^•$	−12.0	14.3	1.1×10^6	2.7×10^6
147	$4-C_8H_{17}NHC_6H_4NHC_8H_{17}$	$HO_2^•$	−22.1	10.0	5.4×10^6	1.0×10^7
147	$4-C_8H_{17}NHC_6H_4NHC_8H_{17}$	*sec*-$RO_2^•$	−18.6	11.4	3.3×10^6	6.5×10^6
147	$4-C_8H_{17}NHC_6H_4NHC_8H_{17}$	*tert*-$RO_2^•$	−11.7	14.4	1.1×10^6	2.6×10^6
148	$4-C_6H_5NHC_6H_4NHC_6H_5$	$HO_2^•$	−13.1	13.8	1.4×10^6	3.2×10^6
148	$4-C_6H_5NHC_6H_4NHC_6H_5$	*sec*-$RO_2^•$	−9.6	15.4	7.7×10^5	1.9×10^6
148	$4-C_6H_5NHC_6H_4NHC_6H_5$	*tert*-$RO_2^•$	−2.7	18.6	2.4×10^5	6.8×10^5
149	$4-(CH_3)_2CHC_6H_4NH-4-C_6H_4NHC_6H_4CH(CH_3)_3$	$HO_2^•$	−35.4	4.7	3.7×10^7	4.9×10^7
149	$4-(CH_3)_2CHC_6H_4NH-4-C_6H_4NHC_6H_4CH(CH_3)_3$	*sec*-$RO_2^•$	−31.9	6.0	2.3×10^7	3.3×10^7
149	$4-(CH_3)_2CHC_6H_4NH-4-C_6H_4NHC_6H_4CH(CH_3)_3$	*tert*-$RO_2^•$	−25.0	8.8	8.3×10^6	1.4×10^7
150	$4-C_6H_5NHC_6H_4NHCH(CH_3)_2$	$HO_2^•$	−19.8	10.9	3.9×10^6	7.5×10^6
150	$4-C_6H_5NHC_6H_4NHCH(CH_3)_2$	*sec*-$RO_2^•$	−16.3	12.4	2.3×10^6	4.8×10^6
150	$4-C_6H_5NHC_6H_4NHCH(CH_3)_2$	*tert*-$RO_2^•$	−9.4	15.5	7.4×10^5	1.9×10^6
151	$C_6H_5NH-1-3,7-[(CH_3)_3C]_2C_{10}H_5$	$HO_2^•$	−24.1	9.1	3.7×10^6	6.5×10^6
151	$C_6H_5NH-1-3,7-[(CH_3)_3C]_2C_{10}H_5$	*sec*-$RO_2^•$	−20.6	10.6	2.2×10^6	4.1×10^6
151	$C_6H_5NH-1-3,7-[(CH_3)_3C]_2C_{10}H_5$	*tert*-$RO_2^•$	−13.7	13.5	7.6×10^5	1.7×10^6
152	$1-C_{10}H_7NHC_6H_5$	$HO_2^•$	−11.9	14.3	5.7×10^5	1.4×10^6
152	$1-C_{10}H_7NHC_6H_5$	*sec*-$RO_2^•$	−8.4	15.9	3.2×10^5	8.4×10^5

No. of AmH	Amine	$RO_2^•$	ΔH/ kJ mol^{-1}	E/ kJ mol^{-1}	k (333 K)/ l mol^{-1} s^{-1}	k (400 K)/ l mol^{-1} s^{-1}
152	$1\text{-}C_{10}H_7NHC_6H_5$	*tert*-$RO_2^•$	−1.5	19.1	1.0×10^5	3.2×10^5
153	$2\text{-}C_{10}H_7NHC_6H_5$	$HO_2^•$	−6.1	17.0	2.2×10^5	6.0×10^5
153	$2\text{-}C_{10}H_7NHC_6H_5$	*sec*-$RO_2^•$	−2.6	18.6	1.2×10^5	3.7×10^5
153	$2\text{-}C_{10}H_7NHC_6H_5$	*tert*-$RO_2^•$	4.3	17.8	1.6×10^5	4.7×10^5
154	$2\text{-}C_{10}H_7NH(4'\text{-}C_6H_5O)C_6H_4$	$HO_2^•$	−19.5	11.0	1.2×10^6	3.7×10^7
154	$2\text{-}C_{10}H_7NH(4'\text{-}C_6H_5O)C_6H_4$	*sec*-$RO_2^•$	−16.0	12.5	1.1×10^6	2.3×10^6
154	$2\text{-}C_{10}H_7NH(4'\text{-}C_6H_5O)C_6H_4$	*tert*-$RO_2^•$	−9.1	15.6	3.6×10^5	9.2×10^5
155	1,2-*cyclo*-[C(*cyclo*-CH_2CH_2O)CH_2C (NH-(*cyclo*-C_6H_{11}))]C_6H_4	$HO_2^•$	0.8	20.2	6.8×10^4	2.3×10^5
155	1,2-*cyclo*-[C(*cyclo*-CH_2CH_2O)CH_2C (NH-(*cyclo*-C_6H_{11}))]C_6H_4	*sec*-$RO_2^•$	4.3	21.9	3.7×10^4	1.4×10^5
155	1,2-*cyclo*-[C(*cyclo*-CH_2CH_2O)CH_2C (NH-(*cyclo*-C_6H_{11}))]C_6H_4	*tert*-$RO_2^•$	11.2	25.4	1.0×10^4	4.8×10^4
156	$1\text{-}[NHC_6H_4\text{-}4'\text{-}C(CH_3)_3]\text{-}4\text{-}(CH_3)_3CC_{10}H_6$	$HO_2^•$	−16.9	12.1	1.3×10^6	2.6×10^6
156	$1\text{-}[NHC_6H_4\text{-}4'\text{-}C(CH_3)_3]\text{-}4\text{-}(CH_3)_3CC_{10}H_6$	*sec*-$RO_2^•$	−13.4	13.7	7.1×10^5	1.6×10^6
156	$1\text{-}[NHC_6H_4\text{-}4'\text{-}C(CH_3)_3]\text{-}4\text{-}(CH_3)_3CC_{10}H_6$	*tert*-$RO_2^•$	−6.5	16.8	2.3×10^5	6.4×10^5
157	1,2-*cyclo*-[$C(CH_3)CH_2C(CH_3)_2NH$]C_6H_4	$HO_2^•$	−0.5	19.6	8.4×10^4	2.8×10^5
157	1,2-*cyclo*-[$C(CH_3)CH_2C(CH_3)_2NH$]C_6H_4	*sec*-$RO_2^•$	3.0	21.3	4.6×10^4	1.7×10^5
157	1,2-*cyclo*-[$C(CH_3)CH_2C(CH_3)_2NH$]C_6H_4	*tert*-$RO_2^•$	9.9	24.8	1.3×10^4	5.8×10^4
	$C_6H_5NH_2$	$HO_2^•$	−0.8	19.5	1.7×10^5	5.7×10^5
	$C_6H_5NH_2$	*sec*-$RO_2^•$	2.7	21.2	9.4×10^4	1.7×10^5
	$C_6H_5NH_2$	*tert*-$RO_2^•$	9.6	24.6	2.8×10^4	1.2×10^5
	$C_6H_5(CH_3)NH$	$HO_2^•$	−2.9	18.5	1.3×10^5	3.8×10^5
	$C_6H_5(CH_3)NH$	*sec*-$RO_2^•$	0.6	20.1	7.0×10^4	2.4×10^5
	$C_6H_5(CH_3)NH$	*tert*-$RO_2^•$	7.5	23.5	2.1×10^4	8.5×10^4
	$(CH_3)_2NH$	$HO_2^•$	13.8	26.8	6.3×10^3	3.2×10^4
	$(CH_3)_2NH$	*sec*-$RO_2^•$	17.3	28.6	3.3×10^3	1.8×10^4
	$(CH_3)_2NH$	*tert*-$RO_2^•$	24.2	32.4	8.3×10^2	5.9×10^3

Table 4.3
Enthalpies, activation energies and rate constants of reaction of aminyl radicals with secondary and tertiary hydroperoxides calculated by formulas 1.15–1.17 and 1.20. The values of A, br_e and α, see Table 1.6

No. of AmH	Aminyl radical	ROOH	ΔH/ kJ mol^{-1}	E/ kJ mol^{-1}	k(333 K)/ l mol^{-1} s^{-1}	k(400 K)/ l mol^{-1} s^{-1}
135	$(2\text{-}C_{10}H_7)_2N^{\bullet}$	*sec*-ROOH	5.3	23.8	1.8×10^4	7.8×10^4
135	$(2\text{-}C_{10}H_7)_2N^{\bullet}$	*tert*-ROOH	−1.6	20.2	6.8×10^4	2.3×10^5
136	$(C_6H_5)_2N^{\bullet}$	*sec*-ROOH	0.8	21.4	4.4×10^4	1.6×10^5
136	$(C_6H_5)_2N^{\bullet}$	*tert*-ROOH	−6.1	17.9	1.6×10^5	4.6×10^5
137	$(4\text{-}BrC_6H_4)_2N^{\bullet}$	*sec*-ROOH	1.3	21.7	3.9×10^4	1.5×10^5
137	$(4\text{-}BrC_6H_4)_2N^{\bullet}$	*tert*-ROOH	5.6	24.0	1.7×10^4	7.3×10^4
138	$(4\text{-}CH_3OC_6H_4)_2N^{\bullet}$	*sec*-ROOH	16.9	30.3	1.8×10^3	1.1×10^4
138	$(4\text{-}CH_3OC_6H_4)_2N^{\bullet}$	*tert*-ROOH	10.0	26.4	7.2×10^3	3.6×10^4
139	$(4\text{-}CH_3C_6H_4)_2N^{\bullet}$	*sec*-ROOH	8.0	25.3	1.1×10^4	5.0×10^4
139	$(4\text{-}CH_3C_6H_4)_2N^{\bullet}$	*tert*-ROOH	1.1	21.6	4.1×10^4	1.5×10^5
140	$(4\text{-}(CH_3)_3CC_6H_4)_2N^{\bullet}$	*sec*-ROOH	6.7	24.6	1.4×10^4	6.1×10^4
140	$(4\text{-}(CH_3)_3CC_6H_4)_2N^{\bullet}$	*tert*-ROOH	−0.2	20.9	5.3×10^4	1.9×10^5
141	$4\text{-}CH_3OC_6H_4(C_6H_5)N^{\bullet}$	*sec*-ROOH	9.6	26.2	7.8×10^3	3.8×10^4
141	$4\text{-}CH_3OC_6H_4(C_6H_5)N^{\bullet}$	*tert*-ROOH	2.7	22.4	3.1×10^4	1.2×10^5
142	$4\text{-}NO_2C_6H_4(C_6H_5)N^{\bullet}$	*sec*-ROOH	− 7.4	17.3	1.9×10^5	5.5×10^5
142	$4\text{-}NO_2C_6H_4(C_6H_5)N^{\bullet}$	*tert*-ROOH	−14.3	14.0	6.4×10^5	1.5×10^6
143	$4\text{-}(CH_3)_3CC_6H_4(C_6H_5)N^{\bullet}$	*sec*-ROOH	5.2	23.8	1.8×10^4	7.8×10^4
143	$4\text{-}(CH_3)_3CC_6H_4(C_6H_5)N^{\bullet}$	*tert*-ROOH	− 1.7	20.2	6.8×10^4	2.3×10^5
144	$1\text{-}C_{10}H_7NH^{\bullet}$	*sec*-ROOH	− 9.2	16.4	2.7×10^5	7.2×10^5
144	$1\text{-}C_{10}H_7NH^{\bullet}$	*tert*-ROOH	−16.1	13.1	8.8×10^5	1.9×10^6
145	$2\text{-}C_{10}H_7NH^{\bullet}$	*sec*-ROOH	−14.0	14.1	6.1×10^5	1.4×10^6
145	$2\text{-}C_{10}H_7NH^{\bullet}$	*tert*-ROOH	−20.9	10.9	1.9×10^6	3.8×10^6
146	$2\text{-}C_{10}H_7NH\text{-}4\text{-}C_6H_4N^{\bullet}\text{-}2'\text{-}C_{10}H_7$	*sec*-ROOH	18.9	31.3	1.2×10^3	8.2×10^3
146	$2\text{-}C_{10}H_7NH\text{-}4\text{-}C_6H_4N^{\bullet}\text{-}2'\text{-}C_{10}H_7$	*tert*-ROOH	12.0	27.5	4.9×10^3	2.6×10^4
147	$4\text{-}C_8H_{17}NHC_6H_4N^{\bullet}C_8H_{17}$	*sec*-ROOH	18.6	31.3	1.2×10^3	8.2×10^3

No. of AmH	Aminyl radical	ROOH	ΔH/ kJ mol^{-1}	E/ kJ mol^{-1}	k(333 K)/ l mol^{-1} s^{-1}	k(400 K)/ l mol^{-1} s^{-1}
147	$4\text{-}C_8H_{17}NHC_6H_4N^{\bullet}C_8H_{17}$	*tert*-ROOH	11.7	27.3	5.2×10^3	2.7×10^4
148	$4\text{-}C_6H_5NHC_6H_4N^{\bullet}C_6H_5$	*sec*-ROOH	9.6	26.2	7.8×10^3	3.8×10^4
148	$4\text{-}C_6H_5NHC_6H_4N^{\bullet}C_6H_5$	*tert*-ROOH	2.7	22.4	3.1×10^4	1.5×10^5
149	$4\text{-}(CH_3)_2CHC_6H_4N^{\bullet}\text{-}4'\text{-}C_6H_4NHC_6H_4CH(CH_3)_3$	*sec*-ROOH	31.9	39.3	0.7	7.4×10^2
149	$4\text{-}(CH_3)_2CHC_6H_4N^{\bullet}\text{-}4'\text{-}C_6H_4NHC_6H_4CH(CH_3)_3$	*tert*-ROOH	25.0	35.1	3.1×10^2	2.6×10^3
150	$4\text{-}C_6H_5N^{\bullet}C_6H_4NHCH(CH_3)_2$	*sec*-ROOH	16.3	30.0	2.0×10^3	1.2×10^4
150	$4\text{-}C_6H_5N^{\bullet}C_6H_4NHCH(CH_3)_2$	*tert*-ROOH	9.4	26.1	8.1×10^3	3.9×10^4
151	$C_6H_5N^{\bullet}\text{-}1,3,7\text{-}[(CH_3)_3C]_2C_{10}H_5$	*sec*-ROOH	20.6	32.5	8.0×10^2	5.7×10^3
151	$C_6H_5N^{\bullet}\text{-}1,3,7\text{-}[(CH_3)_3C]_2C_{10}H_5$	*tert*-ROOH	13.7	28.5	3.4×10^3	1.9×10^4
152	$1\text{-}C_{10}H_7(C_6H_5)N^{\bullet}$	*sec*-ROOH	8.4	25.5	1.0×10^4	4.7×10^4
152	$1\text{-}C_{10}H_7(C_6H_5)N^{\bullet}$	*tert*-ROOH	1.5	21.8	3.8×10^4	1.4×10^5
153	$2\text{-}C_{10}H_7(C_6H_5)N^{\bullet}$	*sec*-ROOH	2.6	22.4	3.1×10^4	1.2×10^5
153	$2\text{-}C_{10}H_7(C_6H_5)N^{\bullet}$	*tert*-ROOH	4.3	23.3	2.2×10^4	9.1×10^4
154	$2\text{-}C_{10}H_7N^{\bullet}(4'\text{-}C_6H_5O)C_6H_4$	*sec*-ROOH	16.0	29.8	2.1×10^3	1.3×10^4
154	$2\text{-}C_{10}H_7N^{\bullet}(4'\text{-}C_6H_5O)C_6H_4$	*tert*-ROOH	9.1	25.9	8.7×10^3	4.1×10^4
155	1,2-*cyclo*-[C(*cyclo*-CH_2CH_2O)CH_2C (N$^{\bullet}$-(*cyclo*-C_6H_{11}))]C_6H_4	*sec*-ROOH	−4.3	18.8	1.1×10^5	3.5×10^5
155	1,2-*cyclo*-[C(*cyclo*-CH_2CH_2O)CH_2C (N$^{\bullet}$-(*cyclo*-C_6H_{11}))]C_6H_4	*tert*-ROOH	−11.2	25.7	9.3×10^3	4.4×10^4
156	$1\text{-}[N^{\bullet}C_6H_4\text{-}4'\text{-}C(CH_3)_3]\text{-}4\text{-}(CH_3)_3CC_{10}H_6$	*sec*-ROOH	13.4	28.3	3.6×10^3	2.0×10^4
156	$1\text{-}[N^{\bullet}C_6H_4\text{-}4'\text{-}C(CH_3)_3]\text{-}4\text{-}(CH_3)_3CC_{10}H_6$	*tert*-ROOH	6.5	24.5	1.4×10^4	6.3×10^4
157	1,2-*cyclo*-[$C(CH_3)CH_2C(CH_3)_2N^{\bullet}$]$C_6H_4$	*sec*-ROOH	3.0	22.6	2.9×10^4	1.1×10^5
157	1,2-*cyclo*-[$C(CH_3)CH_2C(CH_3)_2N^{\bullet}$]$C_6H_4$	*tert*-ROOH	9.9	26.3	7.5×10^3	3.7×10^4
	$C_6H_5N^{\bullet}H$	*sec*-ROOH	−2.7	19.6	8.4×10^4	2.8×10^5
	$C_6H_5N^{\bullet}H$	*tert*-ROOH	−9.6	16.2	2.9×10^5	7.7×10^5
	$C_6H_5(CH_3)N^{\bullet}$	*sec*-ROOH	−0.6	20.7	5.7×10^4	2.0×10^5
	$C_6H_5(CH_3)N^{\bullet}$	*tert*-ROOH	−7.5	17.2	2.0×10^5	5.7×10^5
	$(CH_3)_2N^{\bullet}$	*sec*-ROOH	−17.3	12.6	1.1×10^6	2.3×10^6
	$(CH_3)_2N^{\bullet}$	*tert*-ROOH	−24.2	9.5	3.2×10^6	5.7×10^6

Table 4.4
Enthalpies, activation energies and rate constants of reaction of diphenylaminyl radical with alkylaromatic hydrocarbons calculated by formulas 1.15–1.17 and 1.20. The values of A, br_e and α, see Table 1.6

R_3H	ΔH/ kJ mol^{-1}	E/ kJ mol^{-1}	k(333 K)/ l mol^{-1} s^{-1}	k(400 K)/ l mol^{-1} s^{-1}
$C_6H_5CH_3$	3.5	66.2	1.2×10^{-2}	0.69
4-$CH_3C_6H_4CH_3$	0.7	64.9	4.0×10^{-2}	2.0
4-$CH_3OC_6H_4CH_3$	0.5	64.8	2.0×10^{-2}	1.1
4-$ClC_6H_4CH_3$	2.6	65.7	1.5×10^{-2}	0.78
4-$CNC_6H_4CH_3$	−2.8	63.2	3.6×10^{-2}	1.7
4-$NO_2C_6H_4CH_3$	4.8	66.8	1.0×10^{-2}	0.57
$C_6H_5CH_2CH_3$	−7.7	61.0	5.4×10^{-2}	2.1
$C_6H_5CH(CH_3)_2$	−14.1	58.1	7.2×10^{-2}	2.6
$(C_6H_5)_2CH_2$	−14.7	57.9	0.17	5.5
$(C_6H_5)_2CHCH_3$	−20.5	55.4	0.20	5.8
$C_6H_5CH_2CH_2C_6H_5$	−8.5	60.6	0.12	4.9
$(C_6H_5)_3CH$	−21.9	54.8	0.25	7.0
cyclo-$C_6H_{11}C_6H_5$	−15.1	57.7	8.9×10^{-2}	2.9
Indane	−9.9	60.0	0.16	5.8
Tetraline	−15.7	57.4	0.40	13
1,2-Dihydronaphthalene	−35.0	49.4	3.6	71
1,2-Dihydropyrene	−30.1	51.4	3.5	78
9,10-Dihydrophenanthrene	−27.8	52.3	2.5	59
9,10-Dihydroanthracene	−49.5	43.7	56	7.9×10^{2}

Table 4.5
Enthalpies, activation energies and rate constants of reaction of aminyl radicals with cumene calculated by formulas 1.15–1.17 and 1.20. The values of A, br_e and α, see Table 1.6

No. of AmH	Aminyl radical	ΔH/ kJ mol^{-1}	E/ kJ mol^{-1}	k(333 K)/ l mol^{-1} s^{-1}	k(400 K)/ l mol^{-1} s^{-1}
135	$(2\text{-}C_{10}H_7)_2N^{\bullet}$	−9.6	60.2	3.6×10^{-2}	1.4
136	$(C_6H_5)_2N^{\bullet}$	−14.1	58.2	7.4×10^{-2}	2.5
137	$(4\text{-}BrC_6H_4)_2N^{\bullet}$	−13.6	58.4	6.9×10^{-2}	2.4
138	$(4\text{-}CH_3OC_6H_4)_2N^{\bullet}$	2.0	65.5	5.3×10^{-3}	0.28
139	$(4\text{-}CH_3C_6H_4)_2N^{\bullet}$	−6.9	61.4	2.3×10^{-2}	0.96
140	$(4\text{-}(CH_3)_3CC_6H_4)_2N^{\bullet}$	−8.2	60.8	2.9×10^{-2}	1.1
141	$4\text{-}CH_3OC_6H_4(C_6H_5)N^{\bullet}$	−5.3	62.1	1.8×10^{-2}	0.78
142	$4\text{-}NO_2C_6H_4(C_6H_5)N^{\bullet}$	−22.3	54.6	0.27	7.4
143	$4\text{-}(CH_3)_3CC_6H_4(C_6H_5)N^{\bullet}$	−9.7	60.1	3.7×10^{-2}	1.4
144	$1\text{-}C_{10}H_7NH^{\bullet}$	−24.1	53.8	0.36	9.4
145	$2\text{-}C_{10}H_7NH^{\bullet}$	−28.9	51.8	0.75	17
146	$2\text{-}C_{10}H_7NHC_6H_4N^{\bullet}\text{-}2'\text{-}C_{10}H_7$	4.0	66.4	3.8×10^{-3}	0.24
147	$4\text{-}C_8H_{17}NHC_6H_4N^{\bullet}C_8H_{17}$	3.7	66.2	4.1×10^{-3}	0.23
148	$4\text{-}C_6H_5NHC_6H_4N^{\bullet}C_6H_5$	−5.3	62.1	1.8×10^{-2}	0.78
149	$4\text{-}(CH_3)_2CHC_6H_4N^{\bullet}\text{-}4'\text{-}C_6H_4NHC_6H_4CH(CH_3)_3$	17.0	72.6	4.1×10^{-4}	3.3×10^{-2}
150	$4\text{-}C_6H_5N^{\bullet}C_6H_4NHCH(CH_3)_2$	1.4	65.2	5.9×10^{-3}	0.31
151	$C_6H_5N^{\bullet}\text{-}1,3,7\text{-}[(CH_3)_3C]_2C_{10}H_5$	5.7	67.2	2.9×10^{-3}	0.17
152	$1\text{-}C_{10}H_7(C_6H_5)N^{\bullet}$	−6.5	61.6	2.2×10^{-2}	0.90
153	$2\text{-}C_{10}H_7(C_6H_5)N^{\bullet}$	−12.3	59.0	5.6×10^{-2}	2.0
154	$2\text{-}C_{10}H_7N^{\bullet}(4'\text{-}C_6H_5O)C_6H_4$	1.1	65.0	6.4×10^{-3}	0.32
155	1,2-*cyclo*-[C(*cyclo*-CH_2CH_2O)CH_2C(NH(*cyclo*-C_6H_{11}))]C_6H_4	−19.2	55.9	0.17	5.0
156	$1\text{-}[NHC_6H_4\text{-}4'\text{-}C(CH_3)_3]\text{-}4\text{-}(CH_3)_3CC_{10}H_6$	−1.5	63.8	9.8×10^{-3}	0.47
157	1,2-*cyclo*-[$C(CH_3)CH_2C(CH_3)_2NH$]C_6H_4	−17.9	56.5	0.14	4.2

Table 4.6
Rate constants of recombination and disproportionation of aminyl radicals in hydrocarbon solutions measured by flash photolysis technique. Products and mechanism see in papers[3–10]

No. of AmH	Aminyl radical	T/K	k/ l mol^{-1} s^{-1} or $\log k = A - E/\theta$	Ref.
135	$(2\text{-}C_{10}H_7)_2N^{\bullet}$	298	4.5×10^8	11
136	$(C_6H_5)_2N^{\bullet}$	298	2.7×10^7	11
136	$(C_6H_5)_2N^{\bullet}$	293	1.8×10^7	12
136	$(C_6H_5)_2N^{\bullet}$	298	2.5×10^7	13
137	$(4\text{-}BrC_6H_4)_2N^{\bullet}$	298	6.0×10^7	11
138	$(4\text{-}CH_3OC_6H_4)_2N^{\bullet}$	298	2.3×10^6	11
139	$(4\text{-}CH_3C_6H_4)_2N^{\bullet}$	298	2.3×10^6	11
140	$(4\text{-}(CH_3)_3CC_6H_4)_2N^{\bullet}$	298	3.4×10^6	11
141	$4\text{-}CH_3OC_6H_4(C_6H_5)N^{\bullet}$	298	6.0×10^6	11
143	$4\text{-}(CH_3)_3CC_6H_4(C_6H_5)N^{\bullet}$	298	1.1×10^7	11
146	$2\text{-}C_{10}H_7NH\text{-}4\text{-}C_6H_4N^{\bullet}\text{-}2'\text{-}C_{10}H_7$	298	1.9×10^8	11
148	$4\text{-}C_6H_5NH_2C_6H_4N^{\bullet}C_6H_5$	298	1.6×10^9	11
152	$1\text{-}C_{10}H_7(C_6H_5)N^{\bullet}$	298	6.2×10^8	11
153	$2\text{-}C_{10}H_7(C_6H_5)N^{\bullet}$	298	3.8×10^8	11
	$2\text{-}(CH_3)_3CC_6H_4(C_6H_5)N^{\bullet}$	298	1.1×10^5	11
	$(4\text{-}CH_2C(CH_3)_2CH_2CH(CH_3)_2\text{-}C_6H_4)_2N^{\bullet}$	298	2.4×10^6	11
	$4\text{-}NH_2C_6H_4N^{\bullet}H$	298	4.5×10^8	11
	$4\text{-}N^{\bullet}C_6H_5\text{-}C_6H_4NH\text{-}cyclo\text{-}C_6H_{11}$	298	5.4×10^8	11
	$1\text{-}(4\text{-}NHC_6H_5\text{-}C_6H_4N^{\bullet})\text{-}C_{10}H_7$	298	4.3×10^8	11
	$2\text{-}(4\text{-}NHC_6H_5\text{-}C_6H_4N^{\bullet})\text{-}C_{10}H_7$	298	1.2×10^9	11
	$2,4,6\text{-}[(CH_3)_3C]_3C_6H_2N^{\bullet}H$	208–233	$12.2 - 65.2/\theta$	14

Table 4.7
Rate constants of reaction of *para*-disubstituted diphenylaminyl radicals with phenols in decane estimated by laser photolysis technique[14,15]

No.	Phenol	k(294 K)/ l mol^{-1} s^{-1}			
	para-Substituents	H, H	Br, Br	CH_3, CH_3	CH_3O, H
9	$4\text{-}CH_3C_6H_4OH$	7.0×10^6	9.6×10^6	3.2×10^6	2.2×10^6
26	$4\text{-}HOC_6H_4OH$	1.3×10^7	1.0×10^7	2.8×10^6	2.0×10^6
36	C_6H_5OH	9.8×10^5	7.0×10^5	6.5×10^5	3.6×10^5
51	$2,6\text{-}((CH_3)_3C)_2\text{-}C_6H_3OH$	7.3×10^6	6.4×10^6	5.2×10^6	3.4×10^6
52	$2,4,6\text{-}((CH_3)_3C)_3\text{-}C_6H_2OH$	1.3×10^7	7.9×10^6	1.1×10^7	7.9×10^6
53	$2,6\text{-}((CH_3)_3C)_2\text{-}4\text{-}CH_3CO\text{-}C_6H_2OH$	1.0×10^7	3.4×10^6	6.4×10^6	5.8×10^6
55	$2,6\text{-}((CH_3)_3C)_2\text{-}4\text{-}((CH_3)_3CO)\text{-}C_6H_2OH$	3.5×10^7	2.0×10^7	2.9×10^7	2.6×10^7
58	$2,6\text{-}((CH_3)_3C)_2\text{-}4\text{-}Cl\text{-}C_6H_2OH$	1.4×10^7	9.1×10^6	1.1×10^7	7.3×10^6
60	$2,6\text{-}((CH_3)_3C)_2\text{-}4\text{-}CN\text{-}C_6H_2OH$	1.1×10^7	3.3×10^6	1.2×10^7	9.8×10^6
61	$2,6\text{-}((CH_3)_3C)_2\text{-}4\text{-}CHO\text{-}C_6H_2OH$	1.3×10^7	4.3×10^6	1.1×10^7	9.4×10^6
64	$2,6\text{-}((CH_3)_3C)_2\text{-}4\text{-}CH_3\text{-}C_6H_2OH$	1.2×10^7	7.9×10^6	1.1×10^7	7.5×10^6
78	$4\text{-}CH_3OC_6H_4OH$	1.5×10^8	1.6×10^8	5.9×10^7	2.6×10^7
83	$2,6\text{-}(CH_3)_2C_6H_3OH$	2.1×10^7	1.8×10^7	9.0×10^6	5.1×10^6
110	$2,4,6\text{-}(CH_3)_3C_6H_2OH$	5.0×10^7	4.2×10^7	3.1×10^7	2.1×10^7
	$4\text{-}ClC_6H_4OH$	2.9×10^6	3.8×10^6	1.1×10^6	5.6×10^5
	$2,6\text{-}(CH_3O)_2C_6H_3OH$	2.9×10^7	3.0×10^7	1.7×10^7	1.3×10^7
	$2,5\text{-}((CH_3)_3C)_2\text{-}4\text{-}HOC_6H_2OH$	2.6×10^8	2.5×10^8	2.3×10^8	1.6×10^8
	$2,6\text{-}((CH_3)_3C)_2\text{-}4\text{-}(CH_3)_2NCH_2\text{-}C_6H_2OH$	1.4×10^7	7.8×10^6	8.4×10^7	7.0×10^6
	$2,6\text{-}((CH_3)_3C)_2\text{-}4\text{-}Br\text{-}C_6H_2OH$	1.6×10^7	8.7×10^7	9.0×10^6	7.8×10^6

Table 4.8
Enthalpies, activation energies and rate constants of reaction between diphenylaminyl radical and aromatic amines calculated by formulas 1.15–1.17 and 1.21. The values of A, br_e and α, see Table 1.6

No.	Amine	ΔH/ kJ mol^{-1}	E/ kJ mol^{-1}	k(333 K)/ l mol^{-1} s^{-1}	k(400 K)/ l mol^{-1} s^{-1}
135	$(2\text{-}C_{10}H_7)_2NH$	−4.5	11.6	1.5×10^6	3.1×10^6
136	$(C_6H_5)_2NH$	0.0	13.8	6.8×10^5	1.6×10^6
137	$(4\text{-}BrC_6H_4)_2NH$	−0.5	13.6	7.4×10^5	1.7×10^6
138	$(4\text{-}CH_3OC_6H_4)_2NH$	−16.1	6.2	1.1×10^7	1.6×10^7
139	$(4\text{-}CH_3C_6H_4)_2NH$	−7.2	10.3	2.4×10^6	4.5×10^6
140	$(4\text{-}(CH_3)_3CC_6H_4)_2NH$	−5.9	10.9	1.9×10^6	3.8×10^6
141	$4\text{-}CH_3OC_6H_4NHC_6H_5$	−8.8	9.6	3.1×10^6	5.6×10^6
142	$4\text{-}NO_2C_6H_4NHC_6H_4$	8.2	18.0	1.5×10^5	4.5×10^5
143	$4\text{-}(CH_3)_3CC_6H_4NHC_6H_5$	−4.4	11.6	1.5×10^6	3.1×10^6
144	$1\text{-}C_{10}H_7NH_2$	10.0	19.0	1.0×10^5	3.3×10^5
145	$2\text{-}C_{10}H_7NH_2$	14.8	21.6	4.1×10^4	1.5×10^5
146	$2\text{-}C_{10}H_7NHC_6H_4NH\text{-}2'\text{-}C_{10}H_7$	−18.1	5.4	1.4×10^7	2.0×10^7
147	$4\text{-}C_8H_{17}NHC_6H_4NHC_8H_{17}$	−17.8	5.5	1.4×10^7	1.9×10^7
148	$4\text{-}C_6H_5NHC_6H_4NHC_6H_5$	−8.8	9.6	3.1×10^6	5.6×10^6
149	$4\text{-}(CH_3)_2CHC_6H_4NH\text{-}4\text{-}C_6H_4NHC_6H_4CH(CH_3)_3$	−31.3	0.0	1.0×10^8	1.0×10^8
150	$4\text{-}C_6H_5NHC_6H_4NHCH(CH_3)_2$	−15.5	6.5	9.6×10^6	1.4×10^7
151	$C_6H_5NH\text{-}1\text{-}3,7\text{-}[(CH_3)_3C]_2C_{10}H_5$	−19.8	4.6	1.9×10^7	2.5×10^7
152	$1\text{-}C_{10}H_7NHC_6H_5$	−7.6	10.1	2.6×10^6	4.8×10^6
153	$2\text{-}C_{10}H_7NHC_6H_5$	−1.8	12.9	9.5×10^5	2.1×10^6
154	$2\text{-}C_{10}H_7NH(4'\text{-}C_6H_5O)C_6H_4$	−15.2	6.6	9.2×10^6	1.4×10^7
155	1,2-*cyclo*-[C(*cyclo*-$CH_2CH_2O)CH_2C(NH$-(*cyclo*-C_6H_{11}))]C_6H_4	5.1	16.4	2.7×10^5	7.2×10^5
156	$1\text{-}[NHC_6H_4\text{-}4'\text{-}C(CH_3)_3]\text{-}4\text{-}(CH_3)_3CC_{10}H_6$	−12.6	7.8	6.0×10^6	9.6×10^4
157	1,2-*cyclo*-[$C(CH_3)CH_2C(CH_3)_2NH$]C_6H_4	−3.8	15.7	3.4×10^5	8.9×10^5

Table 4.9
Rate constants of reaction of aromatic amines with oxygen: $AmH + O_2 \rightarrow Am^{\bullet} + HO_2^{\bullet}$ calculated by formula 1.17 with $A = 3 \times 10^{10}$ l mol^{-1} s^{-1} per one N—H-bond and $E = \Delta H$

No.	Amine	E/ kJ mol^{-1}	k(400 K)/ l mol^{-1} s^{-1}	k(500 K)/ l mol^{-1} s^{-1}
135	$(2\text{-}C_{10}H_7)_2NH$	156.8	1.0×10^{-10}	1.2×10^{-6}
136	$(C_6H_5)_2NH$	161.3	2.6×10^{-11}	4.2×10^{-2}
137	$(4\text{-}BrC_6H_4)_2NH$	160.8	3.0×10^{-11}	4.8×10^{-7}
138	$(4\text{-}CH_3OC_6H_4)_2NH$	145.2	3.3×10^{-9}	2.0×10^{-5}
139	$(4\text{-}CH_3C_6H_4)_2NH$	154.1	2.3×10^{-10}	2.4×10^{-6}
140	$(4\text{-}(CH_3)_3CC_6H_4)_2NH$	155.4	1.5×10^{-10}	5.8×10^{-7}
141	$4\text{-}CH_3OC_6H_4NHC_6H_5$	152.5	3.7×10^{-10}	3.5×10^{-6}
142	$4\text{-}NO_2C_6H_4NHC_6H_4$	169.5	2.2×10^{-12}	5.9×10^{-8}
143	$4\text{-}(CH_3)_3CC_6H_4NHC_6H_5$	156.9	9.7×10^{-11}	1.2×10^{-6}
144	$1\text{-}C_{10}H_7NH_2$	171.3	2.6×10^{-12}	7.6×10^{-8}
145	$2\text{-}C_{10}H_7NH_2$	176.1	6.1×10^{-13}	2.4×10^{-8}
146	$2\text{-}C_{10}H_7NHC_6H_4NH\text{-}2'\text{-}C_{10}H_7$	143.2	1.2×10^{-8}	6.6×10^{-5}
147	$4\text{-}C_8H_{17}NHC_6H_4NHC_8H_{17}$	143.5	1.1×10^{-8}	6.1×10^{-5}
148	$4\text{-}C_6H_5NHC_6H_4NHC_6H_5$	152.5	7.3×10^{-10}	7.0×10^{-6}
149	$4\text{-}(CH_3)_2CHC_6H_4NH\text{-}4\text{-}C_6H_4NHC_6H_4CH(CH_3)_3$	130.2	6.0×10^{-7}	1.5×10^{-3}
150	$4\text{-}C_6H_5NHC_6H_4NHCH(CH_3)_2$	145.8	5.5×10^{-9}	3.5×10^{-5}
151	$C_6H_5NH\text{-}1\text{-}3,7\text{-}[(CH_3)_3C]_2C_{10}H_5$	141.5	1.0×10^{-10}	5.0×10^{-5}
152	$1\text{-}C_{10}H_7NHC_6H_5$	153.7	2.5×10^{-10}	2.6×10^{-6}
153	$2\text{-}C_{10}H_7NHC_6H_5$	159.5	4.5×10^{-11}	6.5×10^{-7}
154	$2\text{-}C_{10}H_7NH(4'\text{-}C_6H_5O)C_6H_4$	146.1	2.5×10^{-9}	1.6×10^{-5}
155	1,2-*cyclo*-[C(*cyclo*-$CH_2CH_2O)CH_2C(NH$-(*cyclo*-C_6H_{11}))]C_6H_4	166.4	5.6×10^{-12}	1.2×10^{-7}
156	$1\text{-}[NHC_6H_4\text{-}4'\text{-}C(CH_3)_3]\text{-}4\text{-}(CH_3)_3CC_{10}H_6$	148.7	1.1×10^{-9}	8.8×10^{-6}
157	1,2-*cyclo*-[$C(CH_3)CH_2C(CH_3)_2NH]C_6H_4$	165.1	8.3×10^{-12}	1.7×10^{-7}

Table 4.10
The values of nonstoichiometric coefficients of chain termination by aromatic amines in oxidizing substances RH with cyclic chain termination

No.	Amine	RH	$RO_2^•$	*T* / K	*f*	Ref.
136	$(C_6H_5)_2NH$	1,3-Cyclohexadiene	$HO_2^•$	348	200	17
136	$(C_6H_5)_2NH$	$C_6H_5CH{=}CHCOOC_2H_5$	$HO_2^•$	323	20	18
136	$(C_6H_5)_2NH$	$C_6H_5CH{=}CHCOOCH_3$	$HO_2^•$	323	20	18
136	$(C_6H_5)_2NH$	$C_6H_5CH{=}CHCOOC_6H_5$	$HO_2^•$	323	20	18
136	$(C_6H_5)_2NH$	*cyclo*-$C_6H_{11}OH$	$RO_2^•$	393	56	19
136	$(C_6H_5)_2NH$	$(CH_3)_2N(CH_2)_2OCO(CH_3)C{=}CH_2$	$RO_2^•$	323	10	20
138	$(4\text{-}CH_3OC_6H_4)_2NH$	$(CH_3)_2NCH_2CH_2N(CH_3)_2$	$RO_2^•$	313	26	21
138	$(4\text{-}CH_3OC_6H_4)_2NH$	$(CH_3)_2CHOH$	$RO_2^•$	343	22	22
138	$(4\text{-}CH_3OC_6H_4)_2NH$	*cyclo*-$[N(CH_3)(CH_2)_5]$	$RO_2^•$	323	52	23
138	$(4\text{-}CH_3OC_6H_4)_2NH$	$N(CH_2CH_3)_3$	$RO_2^•$	313	80	23
138	$(4\text{-}CH_3OC_6H_4)_2NH$	*cyclo*-$C_6H_{11}N(CH_3)_2$	$RO_2^•$	323	70	23
138	$(4\text{-}CH_3OC_6H_4)_2NH$	$CH_3CON(CH_2CH_3)_2$	$RO_2^•$	348	30	23
138	$(4\text{-}CH_3OC_6H_4)_2NH$	*cyclo*-$C_6H_{11}NH_2$	$RO_2^•$	348	18	23
138	$(4\text{-}CH_3OC_6H_4)_2NH$	$(CH_3)_2N(CH_2)_2OCO(CH_3)C{=}CH_2$	$RO_2^•$	323	18	20
138	$(4\text{-}CH_3OC_6H_4)_2NH$	$(CH_3)_2N(CH_2)_3CH_3$	$RO_2^•$	323	16	24
138	$(4\text{-}CH_3OC_6H_4)_2NH$	$(CH_3)_2N(CH_2)_2OCOCH_2CH_3$	$RO_2^•$	323	26	24
144	$1\text{-}C_{10}H_7NH_2$	$(CH_3)_2N(CH_2)_2OCO(CH_3)C{=}CH_2$	$RO_2^•$	323	10	20
144	$1\text{-}C_{10}H_7NH_2$	*cyclo*-$C_6H_{11}NH_2$	$RO_2^•$	348	16	25
144	$1\text{-}C_{10}H_7NH_2$	*cyclo*-$C_6H_{11}OH$	$RO_2^•$	393	56	19
144	$1\text{-}C_{10}H_7NH_2$	1,3-Cyclohexadiene	$HO_2^•$	348	28	26
144	$1\text{-}C_{10}H_7NH_2$	*cyclo*-$C_6H_{11}OH$	$RO_2^•$	393	48	17
144	$1\text{-}C_{10}H_7NH_2$	*cyclo*-$C_6H_{11}OH$	$RO_2^•$	413	30	27
144	$1\text{-}C_{10}H_7NH_2$	*cyclo*-$C_6H_{11}OH$	$RO_2^•$	393	28	28
144	$1\text{-}C_{10}H_7NH_2$	*cyclo*-$C_6H_{11}OH$	$RO_2^•$	348	15	29
144	$1\text{-}C_{10}H_7NH_2$	*cyclo*-$C_6H_{11}OH$	$RO_2^•$	348	90	30
144	$1\text{-}C_{10}H_7NH_2$	*cyclo*-$C_6H_{11}OH + H_2O_2$	$HO_2^•$	393	20	17
144	$1\text{-}C_{10}H_7NH_2$	1,3-Cyclohexadiene	$HO_2^•$	348	28	17

No.	Amine	RH	$RO_2^•$	T / K	f	Ref.
144	1-$C_{10}H_7NH_2$	$CH_3(CH_2)_3OH$	$RO_2^•$	347	12	28
144	1-$C_{10}H_7NH_2$	$CH_3(CH_2)_3OH$	$RO_2^•$	383	17	28
144	1-$C_{10}H_7NH_2$	$CH_3(CH_2)_3OH$	$RO_2^•$	347	12	29
144	1-$C_{10}H_7NH_2$	$CH_3CH(OH)CH_2CH_3$	$RO_2^•$	347	12	28
144	1-$C_{10}H_7NH_2$	$(CH_3)_3COH + H_2O_2$	$HO_2^•$	358	9	28
144	1-$C_{10}H_7NH_2$	$(CH_3)_3COH + H_2O_2$	$HO_2^•$	348	22	30
144	1-$C_{10}H_7NH_2$	$(CH_3)_2CHOH$	$RO_2^•$	348	90	30
145	2-$C_{10}H_7NH_2$	$CH_3(CH_2)_3OH$	$RO_2^•$	347	74	30
145	2-$C_{10}H_7NH_2$	*cyclo*-$C_6H_{11}OH$	$RO_2^•$	393	28	19
146	2-$C_{10}H_7NHC_6H_4NH$-2'-$C_{10}H_7$	$(CH_3)_2N(CH_2)_2OCO(CH_3)C{=}CH_2$	$RO_2^•$	323	26	20
146	2-$C_{10}H_7NHC_6H_4NH$-2'-$C_{10}H_7$	1,3-Cyclohexadiene	$HO_2^•$	348	6.5	26
146	2-$C_{10}H_7NHC_6H_4NH$-2'-$C_{10}H_7$	1,3-Cyclohexadiene	$HO_2^•$	313	22	21
146	2-$C_{10}H_7NHC_6H_4NH$-2'-$C_{10}H_7$	*cyclo*-$[N(CH_3)(CH_2)_5]$	$RO_2^•$	323	43	23
146	2-$C_{10}H_7NHC_6H_4NH$-2'-$C_{10}H_7$	*cyclo*-$C_6H_{11}N(CH_3)_2$	$RO_2^•$	323	17	23
146	2-$C_{10}H_7NHC_6H_4NH$-2'-$C_{10}H_7$	$C_6H_5CH{=}CHCOOC_2H_5$	$HO_2^•$	323	40	18
146	2-$C_{10}H_7NHC_6H_4NH$-2'-$C_{10}H_7$	$C_6H_5CH{=}CHC_6H_5$	$HO_2^•$	323	40	18
150	4-$C_6H_5NHC_6H_4NHCH(CH_3)_2$	*cyclo*-$C_6H_{11}OH$	$RO_2^•$	393	200	19
152	1-$C_{10}H_7NHC_6H_5$	*cyclo*-$C_6H_{11}OH$	$RO_2^•$	393	15	19
153	2-$C_{10}H_7NHC_6H_5$	*cyclo*-$C_6H_{11}OH$	$RO_2^•$	393	26	19
	(4-C_6H_5NH-4'-$C_6H_4O)_2Si(CH_3)_2$	$(CH_3)_2CHOH$	$RO_2^•$	343	18	22

REFERENCES

1. **Varlamov, V. T., Denisov, E. T.,** Kinetics of reaction of 2,4,6-tri-tert-butylphenoxyl with aromatic amines in quasistationary regime and N—H bond dissociation energies of aromatic amines, *Izv. Akad Nauk SSSR, Ser. Khim.*, 1990, 743.
2. **Denisov, E. T.,** The ground of high activity of aromatic amines in reaction with $RO_2^•$. Analysis in the scope of parabolic model. *Kinet. Katal.*, 36, N 3,1995.
3. **Musso, H.,** Uber die Zersetzungsprodukte des Tetraphenylhydrazins, *Chem. Ber.*, 92, 2881, 1959.
4. **Waters, W. A., White, J. E.,** Reactions of N-carbazolyl radicals, *J. Chem. Soc. C.*, 1968, 740.
5. **Welzel, P.,** Die thermishe Spaltung des Tetraphenylhydrazines. Ein Beitrag zum Problem der nicht katalisierten Benzidin-Umlagerung, *Chem. Ber.*, 103, 1318, 1970.
6. **Neugebauer, F. A., Fisher, H.,** Aminyl. 4. Zur thermischen Zersetzung *p*-substituiereter Tetraphenylhydrazine, *Chem. Ber.*, 104, 886, 1971.

7. **Neugebauer, F. A., Fischer H., Bamberger, S., Smith, H. O.**, Aminyl. 6. *tert*-Butyl-substituerte 9-Carbazolyl-Radikale, Carbazol-Radikalkationen und Carbazol-9-oxyl-Radical, *Chem. Ber.*, 105, 2644, 1972.
8. **Bridger, R. F.**, Reactions of N-phenyl-2-naphtylamino radicals, *J. Am. Chem. Soc.*, 94, 3124, 1972.
9. **Welzel, P., Gunter, L., Eckhardt, G.**, Die thermische Umlagerung von Tetraarylhydrazinen. Uber die C—N-Verknupfung bei der Bildung von Semidin-Derivaten, *Chem. Ber.*, 107, 3624, 1974.
10. **Welzel, P., Dietz, C., Eckhardt, G.**, Untersuchungen zum Mechanismus der thermischen Umlagerung von Tetraarylhydrazinen. Die Thermolyse einiger deuterierter Tetraarylhydrazine, *Chem. Ber.*, 108, 3550, 1975.
11. **Efremkina, E. A., Khudyakov, I. V., Denisov, E.T.**, The study of recombination and disproportionation of arylaminyl radicals by flash photolysis technique. *Khim. Fiz.*, 6, 1289, 1987.
12. **Varlamov, V. T., Safiullin, R. L., Denisov, E. T.**, The study of selfrecombination of diphenylaminyl radicals and their recombination with peroxyl radicals by flash photolysis technique, *Khim. Fiz.*, 2, 408, 1983.
13. **Shida, T., Kira, A.**, Optical and electron spin resonance studies on photolized and radiolyzed tetraphenylhidrazine and related compounds, *J. Phys. Chem.*, 73, 4315, 1969.
14. **Fisher H., Ed.**, *LANDOLT-BORNSTEIN Numerical Data and Functional Relationships in Science and Technology, New Series*; Springer-Verlag, Berlin, 1984, V. 13, S/vol C
15. **Varlamov, V. T., Denisov, N. N., Nadtochenko, V. A., Marchenko, E. P.**, The study of reaction of diphenylaminyl radicals with *para*-substituted phenols by flash photolysis technique, *Kinet. Katal.*, 35, 833, 1994.
16. **Varlamov, V. T., Denisov, N. N., Nadtochenko, V. A., Marchenko, E. P., Petrov, I. V., Plekhanova, L. G.**, The study of reaction of diphenylaminyl radicals with sterically hindered *para*-substituted phenols by flash photolysis technique, *Kinet. Katal.*, 35, 838, 1994.
17. **Denisov, E. T.**, Regeneration of inhibitors in reaction of chain termination in liquid-phase oxidation, in *Theory and Practice of Liquid-Phase Oxidation*, Emanuel, N. M. Ed., Nauka, Moscow, 1974, 237 (in Russian).
18. **Pliss, E. M.** *Oxidation of Vinyl Compounds:Mechanism, Elementary Steps, Structure and Reactivity Relationship*, Doct. Sci. (Chem.) Thesis Dissertation, Inst. Chem. Phys., Chernogolovka, 1990, 23–24 (in Russian).
19. **Vardanyan, R. L., Kharitonov, V. V., Denisov, E. T.**, Mechanism of retarding action of aromatic amines in oxidizing cyclohexanole and cyclohexene, *Neftekhimiya*, 11, 247, 1971,
20. **Pliss, E. M., Alexandrov, A.L., Mogilevich, M. M.**, Cyclic chain termination by inhibitor molecules in oxidizing dimethylaminoethylmethacrylate, *Izv. Akad Nauk SSSR, Ser. Khim.*, 1977, 1441.
21. **Samatov, U. Ya., Alexandrov, A. L., Akhunov, I. R.**, Oxidation of amines by molecular oxygen. 7. Reactions of peroxyl radicals of N-tetramethylethylenediamine., *Izv. Akad Nauk SSSR, Ser. Khim.*, 1978, 2254.
22. **Denisov, E. T., Goldenberg, V. I., Verba, L.G.**, Mechanism of cyclic chain termination by aromatic amines and their intermediates in oxidizing isopropanole and ethylbenzene, *Izv. Akad Nauk SSSR, Ser. Khim.*, 1988, 2217.
23. **Alexandrov, A. L.**, *Negative Catalysis in Chain Oxidation of Nitrgen and Oxigen-Containing Compounds*, Doct. Sci. (Chem.) Thesis Dissertation, Inst. Chem. Phys., Chernogolovka, 1987, 15 — 17 (in Russian).
24. **Pliss, E. M., Alexandrov, A. L., Mikhlin, V. S., Mogilevich, M.M.**, Oxidation of amines by molecular oxygen. 8. Kinetics of inhibited oxidation of dimethylaminoethylpropionate and dimethylbuthylamine. *Izv. Akad Nauk SSSR, Ser. Khim.*, 1978, 2259.
25. **Kovtun, G. A., Alexandrov, A. L.**, Oxidation of amines by molecular oxygen. 1. Kinetics of inhibited oxidation of primary and secondary amines. *Izv. Akad Nauk SSSR, Ser. Khim.*, 1973, 2208.
26. **Vardanyan, R. L., Denisov, E. T.**, Regeneration of inhibitors in oxidizing 1,3-cyclohexadiene, *Izv. Akad Nauk SSSR, Ser. Khim.*, 1971, 2818.
27. **Denisov, E. T., Kharitonov, V. V.**, Pecularity of inhibiting action of 1-naphthylamine in oxidizing cyclohexanole, *Izv. Akad Nauk SSSR, Ser. Khim.*, 1963, 2222.
28. **Denisov, E. T., Scheredin, V. P.**, Synergistic action of alcohols on inhibiting activity of aromatic amines. *Izv. Akad Nauk SSSR, Ser. Khim.*, 1964, 919.
29. **Kharitonov, V. V., Denisov, E.T.**, The dual reactivity of hydroxyperoxyl radicals in reactins with aromatic amines, *Izv. Akad Nauk SSSR, Ser. Khim.*, 1967, 2764.
30. **Vardanyan, R. L., Kharitonov, V. V., Denisov, E. T.**, Mechanism of 1-naphthylamine regeneration in oxidizing alcohols, , *Izv. Akad Nauk SSSR, Ser. Khim.*, 1970, 1536.

Chapter 5

DISSOCIATION ENERGIES AND RATE CONSTANTS OF REACTIONS OF HYDROXYLAMINES AND NITROXYL RADICALS

Table 5.1
O—H-Bonds dissociation energies of hydroxylamines

No.	Name and formula of hydroxylamine	Mol. wt.	$D_{O—H}$/ kJ mol^{-1}	Ref.
158	9-Azabicyclo[3.3.1]nonanehydroxylamine;	141.22	318.81	1
	cyclo-[-CH(NOH)(CH$_2$)$_3$-CH(CH$_2$)$_3$]			
159	8-Azabicyclo[3.2.1]octanehydroxylamine;	127.19	322.2	1
	cyclo-[-CH(NOH)(CH$_2$)$_2$-CH(CH$_2$)$_3$]			
160	1,4-Diazacycloheptane, 2,2,6,6-tetramethyl-5-oxo;	170.26	304.3	2
	cyclo-[C(CH$_3$)$_2$CH$_2$C(O)NHCH$_2$C(CH$_3$)$_2$N(OH)]			
161	2,5-Dihydroimidazole, 1-hydroxy-2,2,5,5-tetramethyl-4-aceto;	168.24	296.9	3
	cyclo-[N(OH)C(CH$_3$)$_2$C(COCH$_3$)=NC(CH$_3$)$_2$]			
162	2,5-Dihydroimidazole, 1-hydroxy-2,2,5,5-tetramethyl-4'-chlorphenyl;	217.29	294.8	3
	cyclo-[N(OH)C(CH$_3$)$_2$C(4'-ClC$_6$H$_4$)=NC(CH$_3$)$_2$]			
163	2,5-Dihydroimidazole, 1-hydroxy-2,2,5,5-tetramethyl-4-(4'-fluorophenyl);	236.29	296.1	2, 3
	cyclo-[C(CH$_3$)$_2$C(4'-FC$_6$H$_4$)=NC(CH$_3$)$_2$N(OH)]			
164	2,5-Dihydroimidazole, 1-hydroxy-2,2,5,5-tetramethyl-4-formamido;	169.23	296.9	3
	cyclo-[N(OH)C(CH$_3$)$_2$C(CONH$_2$)=NC(CH$_3$)$_2$]			
165	2,5-Dihydroimidazole, 1-hydroxy-2,2,5,5-tetramethyl-4-isopropyl;	184.28	294.9	3
	cyclo-[N(OH)C(CH$_3$)$_2$C(CH(CH$_3$)$_2$)=NC(CH$_3$)$_2$]			
166	2,5-Dihydroimidazole, 1-hydroxy-2,2,5,5-tetramethyl-4-methyl;	156.23	295.2	2, 3
	cyclo-[C(CH$_3$)$_2$C(CH$_3$)=NC(CH$_3$)$_2$N(OH)]			
167	2,5-Dihydroimidazole, 1-hydroxy-2,2,5,5-tetramethyl-4-(4'-methylphenyl);	232.33	296.9	2, 3
	cyclo-[C(CH$_3$)$_2$C(4'-CH$_3$C$_6$H$_4$)=NC(CH$_3$)$_2$N(OH)]			

No.	Name and formula of hydroxylamine	Mol. wt.	D_{O-H}/ kJ mol^{-1}	Ref.
168	2,5-Dihydroimidazole, 1-hydroxy-2,2,5,5-tetramethyl-4-phenyl; *cyclo*-$[C(CH_3)_2C(C_6H_5)=NC(CH_3)_2N(OH)]$	218.30	294.2	2, 3
169	2,5-Dihydroimidazole, 1-hydroxy-2,2,5,5-tetramethyl-4-trichloromethyl; *cyclo*-$[C(CH_3)_2C(CCl_3)=NC(CH_3)_2N(OH)]$	259.57	297.1	2, 3
170	2,5-Dihydroimidazole, 1-hydroxy-2,2,5,5-tetramethyl-4-methylenodimethyamino; *cyclo*-$[NOH)C(CH_3)_2C(CH=N(CH_3)_2)=NC(CH_3)_2]$	196.30	294.9	3
171	2,5-Dihydroimidazole, 1-hydroxy-2,2,5,5-tetramethyl-4-methylenohydroxylamino; *cyclo*-$[N(OH)C(CH_3)_2C(CH=NOH)=NC(CH_3)_2]$	185.23	296.1	3
172	4,5-Dihydroimidazole, 1-hydroxy-2-phenyl-4,4,5,5-tetramethyl; *cyclo*-$[C(CH_3)_2C(CH_3)_2N=C(C_6H_5)N(OH)]$	218.30	312.2	2
173	2,5-Dihydroimidazole, 3-oxide-1-hydroxy-2-methyl-2-methylamino- N′-hydroxy-4-phenyl; *cyclo*-$[C(CH_3)_2C(C_6H_5)=N(O)C(CH_3)[N(CH_3)OH]N(OH)]$	251.31	310.0	2
174	2,5-Dihydroimidazole, 3-oxide-1-hydroxy-2,2,4,5,5-pentamethyl; *cyclo*-$[C(CH_3)_2C(CH_3)=N(O)C(CH_3)_2N(OH)]$	172.23	305.8	2
175	4.5-Dihydroimidazole, 3-oxide-1-hydroxy-2-phenyl-4,4,5,5-tetramethyl; *cyclo*-$[C(CH_3)_2C(CH_3)_2N(O)=C(C_6H_5)N(OH)$;	234.30	297.1	2
176	2,5-Dihydroimidazole, 3-oxide-1-hydroxy-2,2,5,5-tetramethyl-4-(1′-bromoethyl); *cyclo*-$[C(CH_3)_2C(CHBrCH_3)=N(O)C(CH_3)_2N(OH)$	251.14	305.8	2
177	2,5-Dihydroimidazole, 3-oxide-1-hydroxy-2,2,5,5-tetramethyl-4-(4′-bromophenyl); *cyclo*-$[C(CH_3)_2C(4'\text{-}BrC_6H_4)=N(O)C(CH_3)_2N(OH)]$	299.19	306.0	2
178	2,5-Dihydroimidazole, 3-oxide-1-hydroxy-2,2,5,5-tetramethyl-4-dibromomethyl; *cyclo*-$[C(CH_3)_2C(CHBr_2)=N(O)C(CH_3)_2N(OH)]$	316.01	311.4	2
179	2,5-Dihydroimidazole, 3-oxide-1-hydroxy-2,2,5,5-tetramethyl-4-(4′-clorophenyl); *cyclo*-$[C(CH_3)_2C(4'\text{-}ClC_6H_4)=N(O)C(CH_3)_2N(OH)]$	268.75	306.3	2
180	2,5-Dihydroimidazole, 3-oxide-1-hydroxy-2,2,5,5-tetramethyl-4-dichloromethyl; *cyclo*-$[C(CH_3)_2C(CHCl_2)=N(O)C(CH_3)_2N(OH)]$	241.12	309.8	2
181	2,5-Dihydroimidazole, 3-oxide-1-hydroxy-2,2,5,5-tetramethyl-4-(4′-fluorophenyl); *cyclo*-$[C(CH_3)_2C(4'\text{-}FC_6H_4)=N(O)C(CH_3)_2N(OH)]$	252.29	305.8	2

No.	Name and formula of hydroxylamine	Mol. wt.	D_{O-H}/ kJ mol^{-1}	Ref.
182	2,5-Dihydroimidazole, 3-oxide-1-hydroxy-2,5,5-trimethyl-2-methyamino-4-phenyl; *cyclo*-[$C(CH_3)_2C(C_6H_5)=N(O)C(CH_3)(NHCH_3)N(OH)$]	249.32	309.5	2
183	2,5-Dihydroimidazole, 3-oxide-1-hydroxy-2,2,5,5-tetramethyl-4-methybromide; *cyclo*-[$C(CH_3)_2C(CH_2Br)=N(O)C(CH_3)_2N(OH)$]	251.12	306.3	2
184	2,5-Dihydroimidazole, 3-oxide-1-hydroxy-2,2,5,5-tetramethyl-4-methyliodide; *cyclo*-[$C(CH_3)_2C(CH_2I)=N(O)C(CH_3)_2N(OH)$]	298.12	306.8	2
185	2,5-Dihydroimidazole, 3-oxide-1-hydroxy-2,2,5,5-tetramethyl-4-(4′-methoxyphenyl); *cyclo*-[$C(CH_3)_2C(4'\text{-}CH_3OC_6H_4)=N(O)C(CH_3)_2N(OH)$]	252.32	303.2	2
186	2,5-Dihydroimidazole, 3-oxide-1-hydroxy-2,2,5,5-tetramethyl-4-(4′-methylphenyl); *cyclo*-[$C(CH_3)_2C(4'\text{-}CH_3C_6H_4)=N(O)C(CH_3)_2N(OH)$]	236.32	303.6	2
187	2,5-Dihydroimidazole, 3-oxide-1-hydroxy-2,2,5,5-tetramethyl-4-phenyl; *cyclo*-[$C(CH_3)_2C(C_6H_5)=N(O)C(CH_3)_2N(OH)$]	222.29	304.5	2
188	2,5-Dihydropyrrol, 1-hydroxy-2,2,5,5-tetramethyl-3-acetyl; *cyclo*-[$C(CH_3)_2CH=C(COCH_3)C(CH_3)_2N(OH)$]	183.25	294.3	2
189	2,5-Dihydropyrrol, 1-hydroxy-2,2,5,5-tetramethy-3-acetyl-4-bromo; *cyclo*-[$C(CH_3)_2CBr=C(COCH_3)C(CH_3)_2N(OH)$]	262.15	294.5	2
190	2,5-Dihydropyrrol, 1-hydroxy-2,2,5,5-tetramethyl-3-chloro; *cyclo*-[$C(CH_3)_2CH=CClC(CH_3)_2N(OH)$]	175.66	293.3	2
191	2,5-Dihydropyrrol, 1-hydroxy-2,2,5,5-tetramethyl-3,4-dibromo; *cyclo*-[$C(CH_3)_2CBr=CBrC(CH_3)_2N(OH)$]	299.00	293.9	2
192	2,5-Dihydropyrrol, 1-hydroxy-2,2,5,5-tetramethy-3-formamido; *cyclo*-[$C(CH_3)_2CH=C(CONH_2)C(CH_3)_2N(OH)$]	168.24	293.1	2
193	3,4-Dihydropyrrol, 1-hydroxy-2,2,5,5-tetramethyl-3-carboxy; *cyclo*-[$N(OH)C(CH_3)_2C(COOH)=CHC(CH_3)_2$]	169.23	291.2	3
194	3,4-Dihydropyrrol, 1-hydroxy-2,2,5,5-tetramethyl-3-chloro; *cyclo*-$N(OH)C(CH_3)_2CCl=CHC(CH_3)_2$	175.66	292.7	3
195	3,4-Dihydropyrrol, 1-hydroxy-2,2,5,5-tetramethyl-3,4-dibromo; *cyclo*-[$N(OH)C(CH_3)_2CBr=CBrC(CH_3)_2$]	269.00	293.1	3
196	3,4-Dihydropyrrol, 1-hydroxy-2,2,5,5-tetramethyl-3-formamido; *cyclo*-[$N(OH)C(CH_3)_2C(CONH_2)=CHC(CH_3)_2$]	184.24	292.6	3

No.	Name and formula of hydroxylamine	Mol. wt.	D_{O-H}/ kJ mol^{-1}	Ref.
197	3,4-Dihydropyrrol, 1-hydroxy-2,2,5,5-tetramethyl-3-methylcarboxy; *cyclo*-[N(OH)C$(CH_3)_2$C$(COOCH_3)$=CHC$(CH_3)_2$]	199.25	293.4	3
198	3,4-Dihydropyrrol, 1-hydroxy-2,2,5,5-tetramethyl-3-methylcarboxy-4-bromo; *cyclo*-[N(OH)C$(CH_3)_2$C$(COOCH_3)$=CBrC$(CH_3)_2$]	252.13	293.4	3
199	1,2-Dihydroquinoline, 1-hydroxy-2,2,4-trimethyl-6-(ditrifluoromethylhydroxy)methyl; 1,2-C_6H_3[C$(CF_3)_2$OH]-*cyclo*-[C(CH_3)=CHC$(CH_3)_2$N(OH)]	339.29	295.0	4
200	1,2-Dihydroquinoline, 1-hydroxy-2,2,4-trimethyl-6-triphenylmethyl; 1,2-C_6H_3[C$(C_6H_5)_3$]-*cyclo*-[C(CH_3)=CHC$(CH_3)_2$N(OH)]	431.58	289.5	4
201	Hydroxylamine, adamantyl-*tert*-butyl; C_8H_{13}[$(CH_3)_3$C]NOH	197.32	334.7	1
202	Hydroxylamine, benzoyl-*tert*-butyl; C_6H_5C(O)[$(CH_3)_3$C]NOH	193.25	323.0	1
203	Hydroxylamine, N-carboxopiperidine-*tert*-butyl; *cyclo*-[$(CH_2)_5$NC(O)N(OH)C$(CH_3)_3$]	200.28	311.1	5
204	Hydroxylamine, diethyl; $(CH_3CH_2)_2$NOH	89.14	304.4	3
205	Hydroxylamine, di-4-methoxyphenyl; (4-$CH_3OC_6H_4)_2$NOH	229.28	300.4	3
206	Hydroxylamine, di-(4-*tert*-butylphenyl); [4-$(CH_3)_3CC_6H_4]_2$NOH	297.44	296.5	4
207	Hydroxylamine, di-*tert*-butylmethyleno; [$(CH_3)_3C]_2$C=NOH	157.26	338.5	1
208	Hydroxylamine, di-tri-fluoromethyl; $(CF_3)_2$NOH	169.04	345.7	6
209	Hydroxylamine, isopropyl-*tert*-butyl; $(CH_3)_2$CHN(OH)C$(CH_3)_3$	131.22	352.7	1
210	Hydroxylamine, 4-methoxybenzoyl-*tert*-butyl; 4-$CH_3OC_6H_4$C(O)N(OH)C$(CH_3)_3$	223.28	307.7	7
211	Hydroxylamine, 3-methylbutyryl-*tert*-butyl; $(CH_3)_2CHCH_2$C(O)N(OH)C$(CH_3)_3$	173.26	316.5	7

No.	Name and formula of hydroxylamine	Mol. wt.	D_{O-H}/ kJ mol^{-1}	Ref.
212	Hydroxylamine, 4-nitrobenzoyl-*tert*-butyl; $4\text{-}NO_2C_6H_4N(OH)C(CH_3)_3$	210.24	335.6	7
213	Hydroxylamine, 2-phenylacryl-*tert*-butyl; $C_6H_5CH{=}CHC(O)N(OH)C(CH_3)_3$	219.29	321.4	7
214	Hydroxylamine, 4-phenylbenzoyl-*tert*-butyl; $4\text{-}C_6H_5C_6H_4C(O)N(OH)C(CH_3)_3$	253.35	324.4	5
215	Hydroxylamine, 2-phenylpropionyl-*tert*-butyl; $C_6H_5CH_2CH_2C(O)N(OH)C(CH_3)_3$	205.30	318.6	7
216	Hydroxylamine, undecanoyl-*tert*-butyl; $CH_3(CH_2)_9C(O)N(OH)C(CH_3)_3$	241.42	317.2	7
217	Piperidine, 1-hydroxy-2,2,6,6-tetramethyl; *cyclo*-$[C(CH_3)_2(CH_2)_3C(CH_3)_2N(OH)]$	157.26	291.2	1
218	Piperidine, 1-hydroxy-2,2,6,6-tetramethyl-4-benzoyloxy; *cyclo*-$[C(CH_3)_2CH_2CH(OCOC_6H_5)CH_2C(CH_3)_2N(OH)]$	277.37	300.3	2
219	Piperidine, 1-hydroxy-2,2,6,6-tetramethyl-3-bromo-4-oxo; *cyclo*-$[C(CH_3)_2CHBrC(O)CH_2C(CH_3)_2N(OH)]$	250.14	303.4	2
220	Piperidine, 1-hydroxy-2,2,6,6-tetramethyl-3-chloro-4-oxo; *cyclo*-$[C(CH_3)_2CHClC(O)CH_2C(CH_3)_2N(OH)]$	205.69	307.9	2
221	Piperidine, 1-hydroxy-2,2,6,6-tetramethyl-4-hydroxy; *cyclo*-$[C(CH_3)_2CH_2CH(OH)CH_2C(CH_3)_2N(OH)]$	173.26	302.2	2
222	Piperidine, 1-hydroxy-2,2,6,6-tetramethyl-4-oxo; *cyclo*-$[C(CH_3)_2CH_2C(O)CH_2C(CH_3)_2N(OH)]$	171.24	300.4	1
223	Pyrrolidine, 1-hydroxy-2,2,5,5-tetramethyl-3-formamido; *cyclo*-$[N(OH)C(CH_3)_2CH(CONH_2)CH_2C(CH_3)_2]$	186.26	291.0	3
224	Pyrrolidine, 1-hydroxy-2,2,5,5-tetramethyl-3-carboxy; *cyclo*-$[N(OH)C(CH_3)_2CH(COOH)CH_2C(CH_3)_2]$	171.24	290.1	3
225	Pyrrolidine, 1-hydroxy-2,2,5,5-tetramethyl-3-hydroxy; *cyclo*-$[N(OH)C(CH_3)_2CH(OH)CH_2C(CH_3)_2]$	143.23	289.1	3
226	Pyrrolidine, 1-hydroxy-2,2,5,5-tetramethyl-3-oxo; *cyclo*-$[N(OH)C(CH_3)_2COCH_2C(CH_3)_2]$	141.22	298.7	3

No.	Name and formula of hydroxylamine	Mol. wt.	D_{O-H}/ kJ mol^{-1}	Ref.
227	Tetrahydroimidazole, 1,3-dihydroxy-2,2,5,5-tetramethyl-4-oxo; *cyclo*-$[(CH_3)_2C(O)N(OH)C(CH_3)_2N(OH)]$	174.20	300.9	3
228	Tetrahydroimidazole, 1-hydroxy-2,2,3,4,5,5-hexamethyl; *cyclo*-$[C(CH_3)_2CH(CH_3)N(CH_3)C(CH_3)_2N(OH)]$	172.27	290.8	2
229	Tetrahydroimidazole, 1-hydroxy-2,3,4,5,5-pentamethyl-2-cyclohexyl; *cyclo*-$[N(OH)C(CH_3)_2CH(CH_3)N(CH_3)C(CH_3)($*cyclo*-$C_6H_{11})]$	228.38	286.9	3
230	Tetrahydroimidazole, 1-hydroxy-2,2,4,5,5-pentamethyl-4-(4′-fluorophenyl); *cyclo*-$[C(CH_3)_2CH(4'\text{-}FC_6H_4)N(CH_3)C(CH_3)_2N(OH)]$	252.34	292.0	3
231	Tetrahydroimidazole, 1-hydroxy-2,2,3,5,5,-pentamethyl-4-(4′-methylphenyl); *cyclo*-$[C(CH_3)_2CH(4'\text{-}CH_3C_6H_4)N(CH_3)C(CH_3)_2N(OH)]$	236.36	290.7	2
232	1,2,5,6-Tetrahydropyrimidine, 2,2,4,6,6-pentamethyl-3-oxide; *cyclo*-$[C(CH_3)_2N(O){=}C(CH_3)CH_2C(CH_3)_2N(OH)]$	186.26	308.8	2
233	Tetrahydroimidazole, 1-hydroxy-2,2,3,5,5-pentamethyl-4-phenyl; *cyclo*-$[C(CH_3)_2CH(C_6H_5)N(CH_3)C(CH_3)_2N(OH)]$	234.34	292.6	2
234	Tetrahydroimidazole, 1-hydroxy-2,2,5,5-tetramethyl-4-benzoylmethyleno; *cyclo*-$[N(OH)C(CH_3)_2C({=}CHCOC_6H_5)NHC(CH_3)_2]$	260.34	298.6	3
235	1,2,5,6-Tetrahydropiridine, 1-hydroxy-2,2,5,5-tetramethyl-4-phenyl; *cyclo*-$[C(CH_3)_2CH_2C(C_6H_5){=}CHC(CH_3)_2N(OH)]$	231.34	295.9	4
236	Tetrahydropyrrol, 1-hydroxy-2,2,4,4-tetramethyl-3-formamido; *cyclo*-$[C(CH_3)_2CH_2CH(CONH_2)C(CH_3)]_2N(OH)]$	170.26	290.9	2
237	Tetrahydroquinoline, 1-hydroxy-2,2,4-trimethyl-4-phenyl-6-triphenylmethyl; 1,2-$C_6H_3[C(C_6H_5)_3]$-*cyclo*-$[C(CH_3)(C_6H_5)CH_2C(CH_3)_2N(OH)]$	509.69	292.8	4

Table 5.2
Enthalpies, activation energies and rate constants of reactions of secondary alkyl peroxyl radicals with hydroxylamines in hydrocarbon solutions calculated by formulas 1.15–1.17 and 1.21. The values of *A*, br_e and α, see Table 1.6

No.	Hydroxylamine	ΔH/ kJ mol^{-1}	E/ kJ mol^{-1}	k (333 K)/ l mol^{-1} s^{-1}
158	*cyclo*-[-$CH(NOH)(CH_2)_3$-$CH(CH_2)_3$]	−46.7	5.4	4.6×10^6
159	*cyclo*-[-$CH(NOH)(CH_2)_2$-$CH(CH_2)_3$]	−43.3	6.7	2.8×10^6
160	*cyclo*-[$C(CH_3)_2CH_2C(O)NHCH_2C(CH_3)_2N(OH)$]	−61.2	0.30	2.9×10^7
163	*cyclo*-[$C(CH_3)_2C(4'\text{-}FC_6H_4)$=$NC(CH_3)_2N(OH)$]	−68.4	0.00	3.2×10^7
166	*cyclo*-[$C(CH_3)_2C(CH_3)$=$NC(CH_3)_2N(OH)$]	−70.4	0.00	3.2×10^7
167	*cyclo*-[$C(CH_3)_2C(4'\text{-}CH_3C_6H_4)$=$NC(CH_3)_2N(OH)$]	−67.5	0.00	3.2×10^7
172	*cyclo*-[$C(CH_3)_2C(CH_3)_2N$=$C(C_6H_5)N(OH)$]	−53.5	2.9	1.1×10^7
173	*cyclo*-[$C(CH_3)_2C(C_6H_5)$=$N(O)C(CH_3)[N(CH_3)OH]N(OH)$]	−55.5	2.2	1.4×10^7
174	*cyclo*-[$C(CH_3)_2C(CH_3)$=$N(O)C(CH_3)_2N(OH)$]	−59.7	0.80	2.4×10^7
178	*cyclo*-[$C(CH_3)_2C(CHBr_2)$=$N(O)C(CH_3)_2N(OH)$]	−54.1	2.7	1.2×10^7
183	*cyclo*-[$C(CH_3)_2C(CH_2Br)$=$N(O)C(CH_3)_2N(OH)$]	−59.2	0.97	2.3×10^7
185	*cyclo*-[$C(CH_3)_2C(4'\text{-}CH_3OC_6H_4)$=$N(O)C(CH_3)_2N(OH)$]	−62.3	0.00	3.2×10^7
200	1,2-$C_6H_3[C(C_6H_5)_3]$-*cyclo*-[$C(CH_3)$=$CHC(CH_3)_2N(OH)$]	−76.0	0.00	3.2×10^7
201	$C_8H_{13}[(CH_3)_3C]NOH$	−30.8	11.7	4.7×10^5
202	$C_6H_5C(O)[(CH_3)_3C]NOH$	−42.5	7.0	2.6×10^6
207	$[(CH_3)_3C]_2C$=NOH	−27.0	13.3	2.6×10^5
208	$(CF_3)_2NOH$	−19.8	11.9	4.4×10^5
209	$(CH_3)_2CHN(OH)C(CH_3)_3$	−12.8	19.6	2.7×10^4
211	$(CH_3)_2CHCH_2C(O)N(OH)C(CH_3)_3$	−49.0	4.6	6.1×10^6
212	4-$NO_2C_6H_4N(OH)C(CH_3)_3$	−29.9	12.0	4.2×10^5
213	C_6H_5CH=$CHC(O)N(OH)C(CH_3)_3$	−44.1	6.4	3.2×10^6
214	4-$C_6H_5C_6H_4C(O)N(OH)C(CH_3)_3$	−41.1	7.5	2.1×10^6
215	$C_6H_5CH_2CH_2C(O)N(OH)C(CH_3)_3$	−46.9	5.3	4.7×10^6
216	$CH_3(CH_2)_9C(O)N(OH)C(CH_3)_3$	−48.3	4.8	5.7×10^6
217	*cyclo*-[$C(CH_3)_2(CH_2)_3C(CH_3)_2N(OH)$]	−74.3	0.00	3.2×10^7
218	*cyclo*-[$C(CH_3)_2CH_2CH(OC(O)C_6H_5)CH_2C(CH_3)_2N(OH)$]	−65.2	0.00	3.2×10^7

No.	Hydroxylamine	ΔH/ kJ mol^{-1}	E/ kJ mol^{-1}	k (333 K)/ l mol^{-1} s^{-1}
219	*cyclo*-[C(CH$_3$)$_2$CHBrC(O)CH$_2$C(CH$_3$)$_2$N(OH)]	–62.1	0.00	3.2×10^7
220	*cyclo*-[C(CH$_3$)$_2$CHClC(O)CH$_2$C(CH$_3$)$_2$N(OH)]	–67.8	0.00	3.2×10^7
221	*cyclo*-[C(CH$_3$)$_2$CH$_2$CH(OH)CH$_2$C(CH$_3$)$_2$N(OH)]	–63.3	0.00	3.2×10^7
222	*cyclo*-[C(CH$_3$)$_2$CH$_2$C(O)CH$_2$C(CH$_3$)$_2$N(OH)]	–65.1	0.00	3.2×10^7
223	*cyclo*-[C(CH$_3$)$_2$CH(C(O)NH$_2$)CH$_2$C(CH$_3$)$_2$N(OH)]	–74.5	0.00	3.2×10^7
224	*cyclo*-[C(CH$_3$)$_2$CH(C(O)OH)CH$_2$C(CH$_3$)$_2$N(OH)]	–75.4	0.00	3.2×10^7
225	*cyclo*-[C(CH$_3$)$_2$CH(OH)CH$_2$C(CH$_3$)$_2$N(OH)]	–76.4	0.00	3.2×10^7
226	*cyclo*-[C(CH$_3$)$_2$C(O)CH$_2$C(CH$_3$)$_2$N(OH)]	–66.8	0.00	3.2×10^7
227	*cyclo*-[C(CH$_3$)$_2$C(O)N(OH)C(CH$_3$)$_2$N(OH)]	–59.5	0.87	2.3×10^7
232	*cyclo*-[C(CH$_3$)$_2$N(O)=C(CH$_3$)CH$_2$C(CH$_3$)$_2$N(OH)]	–56.7	1.2	6.5×10^7
235	*cyclo*-[C(CH$_3$)$_2$CH$_2$C(C$_6$H$_5$)=CHC(CH$_3$)$_2$N(OH)]	–69.6	0.00	3.2×10^7
237	1,2-C$_6$H$_3$[C(C$_6$H$_5$)$_3$]-*cyclo*-[C(CH$_3$)(C$_6$H$_5$)CH$_2$C(CH$_3$)$_2$N(OH)]	–72.7	0.00	3.2×10^7

Table 5.3
Enthalpies, activation energies and rate constants of reactions of tertiary alkyl peroxyl radicals with hydroxylamines in hydrocarbons solution calculated by formulas 1.15–1.17 and 1.21. The values of A, br_e and α, see Table 1.6

No.	Hydroxylamine	ΔH/ kJ mol^{-1}	E/ kJ mol^{-1}	k(333 K)/ l mol^{-1} s^{-1}
158	*cyclo*-[-CH(NOH)(CH$_2$)$_3$-CH(CH$_2$)$_3$]	–39.8	8.0	1.8×10^6
159	*cyclo*-[-CH(NOH)(CH$_2$)$_2$-CH(CH$_2$)$_3$]	–36.4	9.4	1.1×10^6
160	*cyclo*-[C(CH$_3$)$_2$CH$_2$C(O)NHCH$_2$C(CH$_3$)$_2$N(OH)]	–54.3	2.7	1.2×10^7
163	*cyclo*-[C(CH$_3$)$_2$C(4′-FC$_6$H$_4$)=NC(CH$_3$)$_2$N(OH)]	–61.5	0.2	3.0×10^7
166	*cyclo*-[C(CH$_3$)$_2$C(CH$_3$)=NC(CH$_3$)$_2$N(OH)]	–63.5	0.0	3.2×10^7
167	*cyclo*-[C(CH$_3$)$_2$C(4′-CH$_3$C$_6$H$_4$)=NC(CH$_3$)$_2$N(OH)]	–60.6	0.5	2.7×10^7
172	*cyclo*-[C(CH$_3$)$_2$C(CH$_3$)$_2$N=C(C$_6$H$_5$)N(OH)]	–46.4	5.5	4.4×10^6
173	*cyclo*-[C(CH$_3$)$_2$C(C$_6$H$_5$)=N(O)C(CH$_3$)[N(CH$_3$)OH]N(OH)]	–48.6	4.7	5.9×10^6
174	*cyclo*-[C(CH$_3$)$_2$C(CH$_3$)=N(O)C(CH$_3$)$_2$N(OH)]	–52.8	3.2	1.0×10^7
178	*cyclo*-[C(CH$_3$)$_2$C(CHBr$_2$)=N(O)C(CH$_3$)$_2$N(OH)]	–47.2	5.2	4.9×10^6

No.	Hydroxylamine	ΔH/ kJ mol^{-1}	E/ kJ mol^{-1}	k(333 K)/ l mol^{-1} s^{-1}
183	*cyclo*-[$C(CH_3)_2C(CH_2Br)=N(O)C(CH_3)_2N(OH)$]	−52.3	3.4	9.4×10^6
185	*cyclo*-[$C(CH_3)_2C(4'$-$CH_3OC_6H_4)=N(O)C(CH_3)_2N(OH)$]	−55.4	2.3	1.4×10^7
200	1,2-$C_6H_3[C(C_6H_5)_3]$-*cyclo*-[$C(CH_3)=CHC(CH_3)_2N(OH)$]	−69.1	0.0	3.2×10^7
201	$C_8H_{13}[(CH_3)_3C]NOH$	−23.9	14.6	1.6×10^5
202	$C_6H_5C(O)[(CH_3)_3C]NOH$	−35.6	9.7	9.6×10^5
207	$[(CH_3)_3C]_2C=NOH$	−20.1	16.3	8.9×10^4
208	$(CF_3)_2NOH$	−12.9	19.5	2.8×10^4
209	$(CH_3)_2CHN(OH)C(CH_3)_3$	−5.9	22.9	8.2×10^3
211	$(CH_3)_2CHCH_2C(O)N(OH)C(CH_3)_3$	−42.1	7.1	2.5×10^6
212	4-$NO_2C_6H_4N(OH)C(CH_3)_3$	−23.0	15.0	1.4×10^5
213	$C_6H_5CH=CHC(O)N(OH)C(CH_3)_3$	−37.2	9.1	1.2×10^6
214	4-$C_6H_5C_6H_4C(O)N(OH)C(CH_3)_3$	−34.2	10.3	7.8×10^5
215	$C_6H_5CH_2CH_2C(O)N(OH)C(CH_3)_3$	−40.0	8.0	1.8×10^6
216	$CH_3(CH_2)_9C(O)N(OH)C(CH_3)_3$	−41.4	7.4	2.2×10^6
217	*cyclo*-[$C(CH_3)_2(CH_2)_3C(CH_3)_2N(OH)$]	−67.4	0.0	3.2×10^7
218	*cyclo*-[$C(CH_3)_2CH_2CH(OC(O)C_6H_5)CH_2C(CH_3)_2N(OH)$]	−58.3	1.3	2.0×10^7
219	*cyclo*-[$C(CH_3)_2CHBrC(O)CH_2C(CH_3)_2N(OH)$]	−55.2	2.3	1.4×10^7
220	*cyclo*-[$C(CH_3)_2CHClC(O)CH_2C(CH_3)_2N(OH)$]	−60.9	0.4	2.8×10^7
221	*cyclo*-[$C(CH_3)_2CH_2CH(OH)CH_2C(CH_3)_2N(OH)$]	−56.4	1.9	1.6×10^7
222	*cyclo*-[$C(CH_3)_2CH_2C(O)CH_2C(CH_3)_2N(OH)$]	−58.2	1.3	2.0×10^7
223	*cyclo*-[$C(CH_3)_2CH(C(O)NH_2)CH_2C(CH_3)_2N(OH)$]	−67.6	0.0	3.2×10^7
224	*cyclo*-[$C(CH_3)_2CH(C(O)OH)CH_2C(CH_3)_2N(OH)$]	−68.5	0.0	3.2×10^7
225	*cyclo*-[$C(CH_3)_2CH(OH)CH_2C(CH_3)_2N(OH)$]	−69.5	0.0	3.2×10^7
226	*cyclo*-[$C(CH_3)_2COCH_2C(CH_3)_2N(OH)$]	−59.9	0.7	2.5×10^7
227	*cyclo*-[$C(CH_3)_2C(O)N(OH)C(CH_3)_2N(OH)$]	−52.6	3.3	9.7×10^6
232	*cyclo*-[$C(CH_3)_2N(O)=C(CH_3)CH_2C(CH_3)_2N(OH)$]	−49.8	4.3	6.8×10^6
235	*cyclo*-[$C(CH_3)_2CH_2C(C_6H_5)=CHC(CH_3)_2N(OH)$]	−62.7	0.0	3.2×10^7
237	1,2-$C_6H_3[C(C_6H_5)_3]$-*cyclo*-[$C(CH_3)(C_6H_5)CH_2C(CH_3)_2N(OH)$]	−65.8	0.0	3.2×10^7

Table 5.4
Enthalpies, activation energies and rate constants of hydrogen atom exchange between nitroxyls and hydroxylamines calculated by formulas 1.15–1.17 and 1.21. The values of A, br_e and α, see Table 1.6

No.	Hydroxylamine	ΔH/ kJ mol^{-1}	E/ kJ mol^{-1}	k (333 K)/ l mol^{-1} s^{-1}	k (400 K)/ l mol^{-1} s^{-1}
	$(C_6H_5CO)[(CH_3)_3C]NO^{\bullet}$				
158	*cyclo*-[-CH(NOH)(CH$_2$)$_3$-CH(CH$_2$)$_3$]	−4.2	17.5	5.8×10^4	1.7×10^5
159	*cyclo*-[-CH(NOH)(CH$_2$)$_2$-CH(CH$_2$)$_3$]	−8.0	15.7	1.1×10^5	2.9×10^5
172	*cyclo*-[C(CH$_3$)$_2$C(CH$_3$)$_2$N=C(C$_6$H$_5$)N(OH)]	−10.8	14.4	1.8×10^5	4.2×10^5
173	*cyclo*-[C(CH$_3$)$_2$C(C$_6$H$_5$)=N(O)C(CH$_3$)(N(CH$_3$)OH)N(OH)]	−13.0	13.4	2.5×10^5	5.7×10^5
200	1,2-C$_6$H$_3$[C(C$_6$H$_5$)$_3$]-*cyclo*-[C(CH$_3$)=CHC(CH$_3$)$_2$N(OH)]	−33.5	4.6	6.1×10^6	8.0×10^6
201	C$_8$H$_{13}$[(CH$_3$)$_3$C]NOH	11.7	25.7	3.0×10^3	1.5×10^4
202	C$_6$H$_5$C(O)[(CH$_3$)$_3$C]NOH	0.0	19.6	2.7×10^4	8.8×10^4
207	[(CH$_3$)$_3$C]$_2$C=NOH	15.5	27.7	1.4×10^4	7.7×10^3
208	(CF$_3$)$_2$NOH	22.7	31.7	3.4×10^2	2.3×10^3
209	(CH$_3$)$_2$CHN(OH)C(CH$_3$)$_3$	29.7	35.8	77	6.8×10^2
211	(CH$_3$)$_2$CHCH$_2$C(O)N(OH)C(CH$_3$)$_3$	−6.5	16.4	8.6×10^4	2.3×10^5
213	C$_6$H$_5$CH=CHC(O)N(OH)C(CH$_3$)$_3$	−1.6	18.8	3.6×10^4	1.1×10^5
214	4-C$_6$H$_5$C$_6$H$_4$C(O)N(OH)C(CH$_3$)$_3$	1.4	20.3	2.1×10^4	7.1×10^4
215	C$_6$H$_5$CH$_2$CH$_2$C(O)N(OH)C(CH$_3$)$_3$	−4.4	17.4	6.0×10^4	1.7×10^5
217	*cyclo*-[C(CH$_3$)$_2$(CH$_2$)$_3$C(CH$_3$)$_2$N(OH)]	−31.8	5.3	4.7×10^7	6.5×10^6
218	*cyclo*-[C(CH$_3$)$_2$CH$_2$CH(OCOC$_6$H$_5$)CH$_2$C(CH$_3$)$_2$N(OH)]	−22.7	9.1	1.2×10^6	2.1×10^6
219	*cyclo*-[C(CH$_3$)$_2$CHBrC(O)CH$_2$C(CH$_3$)$_2$N(OH)]	−19.6	10.5	7.2×10^5	1.4×10^6
220	*cyclo*-[C(CH$_3$)$_2$CHClC(O)CH$_2$C(CH$_3$)$_2$N(OH)]	−15.1	12.5	3.5×10^5	7.5×10^7
221	*cyclo*-[C(CH$_3$)$_2$CH$_2$CH(OH)CH$_2$C(CH$_3$)$_2$N(OH)]	−20.8	9.9	9.0×10^5	1.6×10^6
222	*cyclo*-[C(CH$_3$)$_2$CH$_2$C(O)CH$_2$C(CH$_3$)$_2$N(OH)]	−22.6	9.1	1.2×10^6	2.1×10^6
227	*cyclo*-[C(CH$_3$)$_2$C(O)N(OH)C(CH$_3$)$_2$N(OH)]	−17.0	11.6	4.8×10^5	9.8×10^5
232	*cyclo*-[C(CH$_3$)$_2$N(O)=C(CH$_3$)CH$_2$C(CH$_3$)$_2$N(OH)]	−14.2	12.8	3.1×10^5	6.8×10^5
235	*cyclo*-[C(CH$_3$)$_2$CH$_2$C(C$_6$H$_5$)=CHC(CH$_3$)$_2$N(OH)]	−27.1	7.2	2.4×10^6	3.7×10^6
237	1,2-C$_6$H$_3$[C(C$_6$H$_5$)$_3$]-*cyclo*-[C(CH$_3$)(C$_6$H$_5$)CH$_2$C(CH$_3$)$_2$N(OH)]	−30.2	5.9	3.8×10^6	5.4×10^6

No.	Hydroxylamine	ΔH/ kJ mol^{-1}	E/ kJ mol^{-1}	k (333 K)/ l mol^{-1} s^{-1}	k (400 K)/ l mol^{-1} s^{-1}
	$[(CH_3)_2CH][(CH_3)_3C]NO^•$				
158	*cyclo*-$[-CH(NOH)(CH_2)_3-CH(CH_2)_3]$	−33.9	45	6.3×10^6	8.3×10^6
159	*cyclo*-$[-CH(NOH)(CH_2)_2-CH(CH_2)_3]$	−30.5	5.8	3.9×10^6	5.6×10^6
172	*cyclo*-$[C(CH_3)_2C(CH_3)_2N{=}C(C_6H_5)N(OH)]$	−40.5	1.9	1.6×10^7	1.8×10^7
173	*cyclo*-$[C(CH_3)_2C(C_6H_5){=}N(O)C(CH_3)(N(CH_3)OHN(OH)$	−42.7	1.1	2.2×10^7	2.3×10^7
178	*cyclo*-$[C(CH_3)_2C(CHBr_2){=}N(O)C(CH_3)_2N(OH)]$	−41.3	1.7	1.7×10^7	1.9×10^7
200	1,2-$C_6H_3[(C_6H_5)_3]$-*cyclo*-$[C(CH_3){=}CHC(CH_3)_2N(OH)]$	−63.2	0.0	3.2×10^7	3.2×10^7
201	$C_8H_{13}[(CH_3)_3C]NOH$	−18.0	11.1	5.8×10^5	1.1×10^6
202	$C_6H_5C(O)[(CH_3)_3C]NOH$	−29.7	6.1	3.5×10^6	5.1×10^6
203	*cyclo*-$[(CH_2)_5NC(O)N(OH)C(CH_3)_3]$	−41.6	1.6	1.8×10^7	2.0×10^7
207	$[(CH_3)_3C]_2C{=}NOH$	−14.2	12.8	3.1×10^5	6.8×10^5
208	$(CF_3)_2NOH$	−7.0	16.2	9.2×10^4	2.5×10^5
209	$(CH_3)_2CHN(OH)C(CH_3)_3$	0.0	19.6	2.7×10^4	8.8×10^4
210	4-$CH_3OC_6H_4C(O)N(OH)C(CH_3)_3$	−45.0	0.3	2.9×10^7	3.0×10^7
211	$(CH_3)_2CHCH_2C(O)N(OH)C(CH_3)_3$	−36.2	3.6	8.7×10^6	1.1×10^7
212	4-$NO_2C_6H_4N(OH)C(CH_3)_3$	−17.1	11.6	4.8×10^5	9.8×10^5
213	$C_6H_5CH{=}CHC(O)N(OH)C(CH_3)_3$	−31.3	5.5	4.4×10^6	6.1×10^6
214	4-$C_6H_5C_6H_4C(O)N(OH)C(CH_3)_3$	−28.3	6.7	2.8×10^6	4.3×10^6
215	$C_6H_5CH_2CH_2C(O)N(OH)C(CH_3)_3$	−34.1	4.4	6.5×10^6	8.5×10^6
216	$CH_3(CH_2)_9C(O)N(OH)C(CH_3)_3$	−35.5	3.9	7.8×10^6	9.9×10^6
220	*cyclo*-$[C(CH_3)_2CHClC(O)CH_2C(CH_3)_2N(OH)]$	−44.8	0.4	2.7×10^7	2.8×10^7
221	*cyclo*-$[C(CH_3)_2CH_2CH(OH)CH_2C(CH_3)_2N(OH)]$	−50.5	0.0	3.2×10^7	3.2×10^7
222	*cyclo*-$[C(CH_3)_2CH_2C(O)CH_2C(CH_3)_2N(OH)]$	−52.3	0.0	3.2×10^7	3.2×10^7
227	*cyclo*-$[C(CH_3)_2C(O)N(OH)C(CH_3)_2N(OH)]$	−46.7	0.0	3.2×10^7	3.2×10^7
232	*cyclo*-$[C(CH_3)_2N(O){=}C(CH_3)CH_2C(CH_3)_2N(OH)]$	−43.9	0.7	2.5×10^7	2.6×10^7
235	*cyclo*-$[C(CH_3)_2CH_2C(C_6H_5){=}CHC(CH_3)_2N(OH)]$	−56.8	0.0	3.2×10^7	3.2×10^7
237	1,2-$C_6H_3[C(C_6H_5)_3]$-*cyclo*-$[C(CH_3)(C_6H_5)CH_2C(CH_3)_2N(OH)]$	−59.9	0.0	3.2×10^7	3.2×10^7

No.	Hydroxylamine	ΔH/ kJ mol^{-1}	E/ kJ mol^{-1}	k (333 K)/ l mol^{-1} s^{-1}	k (400 K)/ l mol^{-1} s^{-1}
	cyclo-$[C(CH_3)_2(CH_2)_3C(CH_3)_2NO^{\bullet}]$				
158	*cyclo*-$[-CH(NOH)(CH_2)_3-CH(CH_2)_3]$	21.1	30.8	4.7×10^2	3.0×10^3
159	*cyclo*-$[-CH(NOH)(CH_2)_2-CH(CH_2)_3]$	24.5	32.8	2.3×10^2	1.7×10^3
172	*cyclo*-$[-C(CH_3)_2C(CH_3)_2N{=}C(C_6H_5)N(OH)]$	14.5	27.2	1.7×10^3	9.0×10^3
173	*cyclo*-$[C(CH_3)_2C(C_6H_5){=}N(O)C(CH_3)(N(CH_3)OH(N(OH)]$	12.3	26.0	2.7×10^3	1.3×10^4
200	1,2-$C_6H_3[C(C_6H_5)_3]$-*cyclo*-$[C(CH_3){=}CHC(CH_3)_2N(OH)]$	–8.2	15.6	1.1×10^5	2.9×10^5
201	$C_8H_{13}[(CH_3)_3C]NOH$	37.0	40.2	16	1.8×10^2
202	$C_6H_5C(O)[(CH_3)_3C]NOH$	25.3	33.2	2.0×10^2	1.5×10^3
207	$[(CH_3)_3C{=}2NOH$	40.8	42.6	6.6	88
208	$(CF_3)_2NOH$	48.0	48.0	0.95	17
209	$(CH_3)_2CHN(OH)C(CH_3)_3$	55.0	55.0	0.075	2.1
211	$(CH_3)_2CHCH_2C(O)N(OH)C(CH_3)_3$	18.8	29.5	7.7×10^2	4.4×10^3
213	$C_6H_5CH{=}CHC(O)N(OH)C(CH_3)_3$	23.7	32.3	2.7×10^2	1.9×10^3
214	4-$C_6H_5C_6H_4C(O)N(OH)C(CH_3)_3$	26.7	34.0	1.5×10^2	1.1×10^3
215	$C_6H_5CH_2CH_2C(O)N(OH)C(CH_3)_3$	20.9	30.7	4.9×10^2	3.1×10^3
217	*cyclo*-$[C(CH_3)_2(CH_2)_3C(CH_3)_2N(OH)]$	0.0	19.7	2.6×10^4	8.6×10^4
219	*cyclo*-$[C(CH_3)_2CHBrC(O)CH_2C(CH_3)_2N(OH)]$	12.2	26.0	2.7×10^3	1.3×10^4
220	*cyclo*-$[C(CH_3)_2CHClC(O)CH_2C(CH_3)_2N(OH)]$	16.7	28.4	1.1×10^3	6.3×10^3
221	*cyclo*-$[C(CH_3)_2CH_2CH(OH)CH_2C(CH_3)_2N(OH)]$	4.5	21.9	1.2×10^4	4.4×10^4
222	*cyclo*-$[C(CH_3)_2CH_2C(O)CH_2C(CH_3)_2N(OH)]$	2.7	21.0	1.6×10^4	5.8×10^4
224	*cyclo*-$[N(OH)C(CH_3)_2CH(COOH)CH_2C(CH_3)_2]$	–1.1	19.1	3.2×10^4	1.0×10^5
225	*cyclo*-$[N(OH)C(CH_3)_2CH(OH)CH_2C(CH_3)_2]$	–2.1	18.6	3.9×10^4	1.2×10^5
227	*cyclo*-$[C(CH_3)_2C(O)N(OH)C(CH_3)_2N(OH)]$	8.3	23.8	5.9×10^3	2.5×10^4
232	*cyclo*-$[C(CH_3)_2N(O){=}C(CH_3)CH_2C(CH_3)_2N(OH)$	11.1	25.3	3.4×10^3	1.6×10^4
235	*cyclo*-$[C(CH_3)_2CH_2C(C_6H_5){=}CHC(CH_3)_2N(OH)]$	–1.8	18.7	3.7×10^4	1.2×10^5
237	1,2-$C_6H_3[C(C_6H_5)_3]$-*cyclo*-$[C(CH_3)(C_6H_5)CH_2C(CH_3)_2N(OH)]$	–4.9	17.2	6.4×10^4	1.8×10^5

No.	Hydroxylamine	ΔH/ kJ mol^{-1}	E/ kJ mol^{-1}	k (333 K)/ l mol^{-1} s^{-1}	k (400 K)/ l mol^{-1} s^{-1}
	cyclo-[C(CH$_3$)$_2$C(CH$_3$)=N(O)C(CH$_3$)$_2$NO$^•$]				
158	*cyclo*-[-CH(NOH)(CH$_2$)$_3$-CH(CH$_2$)$_3$]	13.0	26.4	2.3×10^3	1.1×10^4
159	*cyclo*-[-CH(NOH)(CH$_2$)$_2$-CH(CH$_2$)$_3$]	16.4	28.2	1.2×10^3	6.6×10^3
172	*cyclo*-[C(CH$_3$)$_2$C(CH$_3$)$_2$N=C(C$_6$H$_5$)N(OH)]	6.4	22.9	8.2×10^3	3.3×10^4
173	*cyclo*-[C(CH$_3$)$_2$C(C$_6$H$_5$)=N(O)C(CH$_3$)(N(CH$_3$)OH)N(OH)]	4.2	21.7	1.3×10^4	4.7×10^4
200	1,2-C$_6$H$_3$[C(C$_6$H$_5$)$_3$]-*cyclo*-[C(CH$_3$)=CHC(CH$_3$)$_2$N(OH)]	−16.3	11.9	4.3×10^5	8.9×10^5
201	C$_8$H$_{13}$[(CH$_3$)$_3$C]NOH	28.9	35.3	93	7.9×10^2
202	C$_6$H$_5$C(O)[(CH$_3$)$_3$C]NOH	17.2	28.6	1.0×10^3	5.9×10^3
207	[(CH$_3$)$_3$C]$_2$C=NOH	32.7	37.6	40	3.9×10^2
208	(CF$_3$)$_2$NOH	39.9	42.0	8.2	1.0×10^2
209	(CH$_3$)$_2$CHN(OH)C(CH$_3$)$_3$	46.9	46.5	1.6	27
211	(CH$_3$)$_2$CHCH$_2$C(O)N(OH)C(CH$_3$)$_3$	10.7	25.1	3.7×10^3	1.7×10^4
213	C$_6$H$_5$CH=CHC(O)N(OH)C(CH$_3$)$_3$	15.6	27.8	1.4×10^3	7.5×10^3
214	4-C$_6$H$_5$C$_6$H$_4$C(O)N(OH)C(CH$_3$)$_3$	18.6	29.4	7.8×10^2	4.6×10^3
215	C$_6$H$_5$CH$_2$CH$_2$C(O)N(OH)C(CH$_3$)$_3$	12.8	26.2	2.5×10^3	1.2×10^4
217	*cyclo*-[C(CH$_3$)$_2$(CH$_2$)$_3$C(CH$_3$)$_2$N(OH)]	−14.6	12.7	3.3×10^5	7.0×10^5
219	*cyclo*-[C(CH$_3$)$_2$CHBrC(O)CH$_2$C(CH$_3$)$_2$N(OH)]	−2.4	18.5	4.0×10^4	1.2×10^5
220	*cyclo*-[C(CH$_3$)$_2$CHClC(O)CH$_2$C(CH$_3$)$_2$N(OH)]	2.1	20.7	1.8×10^4	6.3×10^4
221	*cyclo*-[C(CH$_3$)$_2$CH$_2$CH(OH)CH$_2$C(CH$_3$)$_2$N(OH)]	−3.6	17.8	5.2×10^4	1.5×10^5
222	*cyclo*-[C(CH$_3$)$_2$CH$_2$C(O)CH$_2$C(CH$_3$)$_2$N(OH)]	−5.4	16.9	7.1×10^4	2.0×10^5
224	*cyclo*-[N(OH)C(CH$_3$)$_2$CH(COOH)CH$_2$C(CH$_3$)$_2$]	−15.7	12.2	3.9×10^5	8.2×10^5
227	*cyclo*-[C(CH$_3$)$_2$C(O)N(OH)C(CH$_3$)$_2$N(OH)]	0.2	19.7	2.6×10^4	8.6×10^4
232	*cyclo*-[C(CH$_3$)$_2$N(O)=C(CH$_3$)CH$_2$C(CH$_3$)$_2$N(OH)]	3.0	21.1	1.6×10^4	5.6×10^4
235	*cyclo*-[C(CH$_3$)$_2$CH$_2$C(C$_6$H$_5$)=CHC(CH$_3$)$_2$N(OH)]	−9.9	14.8	1.5×10^5	3.7×10^5
237	1,2-C$_6$H$_3$[C(C$_6$H$_5$)$_3$]-*cyclo*-[C(CH$_3$)(C$_6$H$_5$)CH$_2$C(CH$_3$)$_2$N(OH)]	−13.0	13.4	2.5×10^5	5.7×10^5

Table 5.5
Enthalpies, activation energies and rate constants of reactions of phenoxyl radicals with hydroxylamines in nonpolar solvents calculated by formulas 1.15–1.17 and 1.21. The values of A, br_e and α, see Table 1.6

No.	Hydroxylamine	ΔH/ kJ mol^{-1}	E/ kJ mol^{-1}	k (333 K)/ l mol^{-1} s^{-1}	k (400 K)/ l mol^{-1} s^{-1}
	$4\text{-}CH_3C_6H_4O^{\bullet}$				
158	*cyclo*-$[\text{-}CH(NOH)(CH_2)_3\text{-}CH(CH_2)_3]$	−39.0	4.5	2.0×10^7	2.6×10^7
159	*cyclo*-$[\text{-}CH(NOH)(CH_2)_2\text{-}CH(CH_2)_3]$	−35.6	5.8	1.2×10^7	1.7×10^7
172	*cyclo*-$[C(CH_3)_2C(CH_3)_2N{=}C(C_6H_5)N(OH)]$	−45.6	2.1	4.7×10^7	5.3×10^7
173	*cyclo*-$[C(CH_3)_2C(C_6H_5){=}N(O)C(CH_3)[N(CH_3)OH]N(OH)]$	−47.8	1.3	6.3×10^7	6.8×10^7
200	$1,2\text{-}C_6H_3[C(C_6H_5)_3]$-*cyclo*-$[C(CH_3){=}CHC(CH_3)_2N(OH)]$	−68.3	0.0	1.0×10^8	1.0×10^8
201	$C_8H_{13}[(CH_3)_3C]NOH$	−23.1	11.0	1.9×10^6	3.7×10^6
202	$C_6H_5C(O)[(CH_3)_3C]NOH$	−34.8	6.2	1.1×10^7	1.6×10^7
203	*cyclo*-$[(CH_2)_5NC(O)N(OH)C(CH_3)_3]$	−46.7	1.6	5.6×10^7	6.2×10^7
204	$(CH_3CH_2)_2NOH$	−53.4	0.0	1.0×10^{88}	1.0×10^8
207	$[(CH_3)_3C]_2C{=}NOH$	−19.3	12.7	1.0×10^6	2.2×10^6
208	$(CF_3)_2NOH$	−12.1	15.9	3.2×10^5	8.4×10^5
209	$(CH_3)_2CHN(OH)C(CH_3)_3$	−5.1	19.2	9.7×10^4	3.1×10^5
211	$(CH_3)_2CHCH_2C(O)N(OH)C(CH_3)_3$	−41.3	3.6	2.7×10^7	3.4×10^7
213	$C_6H_5CH{=}CHC(O)N(OH)C(CH_3)_3$	−36.4	5.5	1.4×10^7	1.9×10^7
214	$4\text{-}C_6H_5C_6H_4C(O)N(OH)C(CH_3)_3$	−33.4	6.7	8.9×10^6	1.3×10^7
215	$C_6H_5CH_2CH_2C(O)N(OH)C(CH_3)_3$	−39.2	4.4	2.0×10^7	2.7×10^7
220	*cyclo*-$[C(CH_3)_2CHClC(O)CH_2C(CH_3)_2N(OH)]$	−49.9	0.0	1.0×10^8	1.0×10^8
221	*cyclo*-$[C(CH_3)_2CH_2CH(OH)CH_2C(CH_3)_2N(OH)]$	−55.6	0.0	1.0×10^8	1.0×10^8
222	*cyclo*-$[C(CH_3)_2CH_2C(O)CH_2C(CH_3)_2N(OH)]$	−57.4	0.0	1.0×10^8	1.0×10^8
227	*cyclo*-$[C(CH_3)_2C(O)N(OH)C(CH_3)_2N(OH)]$	−51.8	0.0	1.0×10^8	1.0×10^8
232	*cyclo*-$[C(CH_3)_2N(O){=}C(CH_3)CH_2C(CH_3)_2N(OH)]$	−49.0	0.8	7.5×10^7	7.9×10^7
235	*cyclo*-$[C(CH_3)_2CH_2C(C_6H_5){=}CHC(CH_3)_2N(OH)]$	−61.9	0.0	1.0×10^8	1.0×10^8
237	$1,2\text{-}C_6H_3[C(C_6H_5)_3]$-*cyclo*-$[C(CH_3)(C_6H_5)CH_2C(CH_3)_2N(OH)]$	−65.0	0.0	1.0×10^8	1.0×10^8

No.	Hydroxylamine	ΔH/ kJ mol^{-1}	E/ kJ mol^{-1}	k (333 K)/ l mol^{-1} s^{-1}	k (400 K)/ l mol^{-1} s^{-1}
	$4\text{-}(CH_3)_3COC_6H_4O^{\bullet}$				
158	*cyclo*-$[CH(NOH)(CH_2)_3\text{-}CH(CH_2)_3]$	−27.0	9.3	3.4×10^6	6.2×10^6
159	*cyclo*-$[CH(NOH)(CH_2)_2.CH(CH_2)_3]$	−23.6	10.8	2.0×10^6	3.8×10^6
172	*cyclo*-$[C(CH_3)_2C(CH_3)_2N{=}C(C_6H_5)N(OH)]$	−33.6	6.6	9.4×10^6	1.4×10^7
173	*cyclo*-$[C(CH_3)_2C(C_6H_5){=}N(O)C(CH_3)(N(CH_3)OH)N(OH)]$	−35.8	5.8	1.2×10^7	1.8×10^7
200	$1,2\text{-}C_6H_3[C(C_6H_5)_3]$-*cyclo*-$[C(CH_3){=}CHC(CH_3)_2N(OH)]$	−56.3	0.0	1.0×10^8	1.0×10^8
201	$C_8H_{13}[(CH_3)_3C]NOH$	−11.1	16.4	2.7×10^5	7.2×10^5
202	$C_6H_5C(O)[(CH_3)_3C]NOH$	−22.8	11.1	1.8×10^6	3.4×10^6
207	$[(CH_3)_3C]_2C{=}NOH$	−7.3	18.2	1.4×10^5	4.1×10^5
208	$(CF_3)_2NOH$	−0.1	21.7	4.1×10^4	1.5×10^5
209	$(CH_3)_2CHN(OH)C(CH_3)_3$	6.9	25.3	1.1×10^4	5.0×10^4
211	$(CH_3)_2CHCH_2C(O)N(OH)C(CH_3)_3$	−29.3	8.4	4.7×10^6	8.1×10^6
213	$C_6H_5CH{=}CHC(O)N(OH)C(CH_3)_3$	−24.4	10.4	1.5×10^6	4.4×10^6
214	$4\text{-}C_6H_5C_6H_4C(O)N(OH)C(CH_3)_3$	−21.4	11.7	1.5×10^6	2.7×10^6
215	$C_6H_5CH_2CH_2C(O)N(OH)C(CH_3)_3$	−27.2	9.3	3.4×10^6	6.2×10^6
217	*cyclo*-$[C(CH_3)_2(CH_2)_3C(CH_3)_2N(OH)]$	−54.3	0.0	1.0×10^8	1.0×10^8
218	*cyclo*-$[C(CH_3)_2CH_2CH(OCOC_6H_5)CH_2C(CH_3)_2N(OH)]$	−45.5	2.0	4.9×10^7	5.5×10^7
220	*cyclo*-$[C(CH_3)_2CHClC(O)CH_2C(CH_3)_2N(OH)]$	−37.9	4.9	1.7×10^7	2.3×10^7
221	*cyclo*-$[C(CH_3)_2CH_2CH(OH)CH_2C(CH_3)_2N(OH)]$	−43.6	2.8	3.8×10^7	4.4×10^7
222	*cyclo*-$[C(CH_3)_2CH_2C(O)CH_2C(CH_3)_2N(OH)]$	−45.4	2.1	4.7×10^7	5.3×10^7
227	*cyclo*-$[C(CH_3)_2C(O)N(OH)C(CH_3)_2N(OH)]$	−39.8	4.2	2.2×10^7	2.8×10^7
232	*cyclo*-$[C(CH_3)_2N(O){=}C(CH_3)CH_2C(CH_3)_2N(OH)]$	−37.0	5.3	1.5×10^7	2.0×10^7
235	*cyclo*-$[C(CH_3)_2CH_2C(C_6H_5){=}CHC(CH_3)_2N(OH)]$	−49.9	0.5	8.4×10^7	1.0×10^8
237	$1,2\text{-}C_6H_3[C(C_6H_5)_3]$-*cyclo*-$[C(CH_3)(C_6H_5)CH_2C(CH_3)_2N(OH)]$	−53.0	0.0	1.0×10^8	1.0×10^8

No.	Hydroxylamine	ΔH/ kJ mol^{-1}	E/ kJ mol^{-1}	k (333 K)/ l mol^{-1} s^{-1}	k (400 K)/ l mol^{-1} s^{-1}
	α-$(CH_3)_3C_6(O^{\bullet})(CH_2)_2OC(CH_3)[(CH_2)_3CH(CH_3)]_3CH_3$				
158	*cyclo*-$[CH(NOH)(CH_2)_3$-$CH(CH_2)_3]$	−8.9	17.4	1.9×10^5	5.3×10^5
159	*cyclo*-$[CH(NOH)(CH_2)_2$-$CH(CH_2)_3]$	−5.5	19.1	1.0×10^5	3.1×10^5
172	*cyclo*-$[C(CH_3)_2C(CH_3)_2N{=}C(C_6H_5)N(OH)]$	−15.5	14.4	5.6×10^5	1.3×10^6
173	*cyclo*-$[C(CH_3)_2C(C_6H_5){=}N(O)C(CH_3)(N(CH_3)OH)N(OH)]$	−17.7	13.4	7.8×10^5	1.8×10^6
200	1,2-$C_6H_3[C(C_6H_5)_3]$-*cyclo*-$[C(CH_3){=}CHC(CH_3)_2N(OH)]$	−38.2	4.8	1.8×10^7	2.4×10^7
201	$C_8H_{13}[(CH_3)_3C]NOH$	7.0	25.3	1.1×10^4	5.0×10^4
202	$C_6H_5C(O)[(CH_3)_3C]NOH$	−4.7	19.4	9.1×10^4	2.9×10^5
207	$[(CH_3)_3C]_2C{=}NOH$	10.8	27.3	5.3×10^3	2.7×10^4
208	$(CF_3)_2NOH$	18.0	31.3	1.2×10^3	8.1×10^3
209	$(CH_3)_2CHN(OH)C(CH_3)_3$	25.0	35.2	3.0×10^2	2.5×10^3
211	$(CH_3)_2CHCH_2C(O)N(OH)C(CH_3)_3$	−11.2	27.6	4.7×10^3	2.5×10^4
213	$C_6H_5CH{=}CHC(O)N(OH)C(CH_3)_3$	−6.3	18.7	1.2×10^5	3.8×10^5
214	4-$C_6H_5C_6H_4C(O)N(OH)C(CH_3)_3$	−3.3	20.1	6.9×10^4	2.4×10^5
215	$C_6H_5CH_2CH_2C(O)N(OH)C(CH_3)_3$	−9.1	17.3	1.9×10^5	5.6×10^5
217	*cyclo*-$[C(CH_3)_2(CH_2)_3C(CH_3)_2N(OH)]$	−36.5	5.4	1.4×10^7	2.0×10^7
218	*cyclo*-$[C(CH_3)_2CH_2CH(OCOC_6H_5)CH_2C(CH_3)_2N(OH)]$	−27.4	9.1	3.7×10^6	6.5×10^6
220	*cyclo*-$[C(CH_3)_2CHClC(O)CH_2C(CH_3)_2N(OH)]$	−19.8	12.4	1.1×10^6	2.4×10^6
221	*cyclo*-$[C(CH_3)_2CH_2CH(OH)CH_2C(CH_3)_2N(OH)]$	−25.5	10.0	2.7×10^6	5.0×10^6
222	*cyclo*-$[C(CH_3)_2CH_2C(O)CH_2C(CH_3)_2N(OH)]$	−27.3	9.2	3.4×10^6	6.2×10^6
227	*cyclo*-$[C(CH_3)_2C(O)N(OH)C(CH_3)_2N(OH)]$	−21.7	11.6	1.5×10^5	3.1×10^6
232	*cyclo*-$[C(CH_3)_2N(O){=}C(CH_3)CH_2C(CH_3)_2N(OH)]$	−18.9	12.8	9.7×10^5	2.1×10^6
235	*cyclo*-$[C(CH_3)_2CH_2C(C_6H_5){=}CHC(CH_3)_2N(OH)]$	−31.8	7.4	6.9×10^6	7.2×10^3
237	1,2-$C_6H_3[C(C_6H_5)_3]$-*cyclo*-$[C(CH_3)(C_6H_5)CH_2C(CH_3)_2N(OH)]$	−34.9	6.1	1.1×10^7	2.8×10^3

Table 5.6

Enthalpies, activation energies and rate constants of reactions of sterically hindered phenoxyls with hydroxylamines in nonpolar solutions calculated by formulas 1.15–1.17 and 1.21. The values of A, br_e and α, see Table 1.6

No.	Hydroxylamine	ΔH/ kJ mol^{-1}	E/ kJ mol^{-1}	k (333 K)/ l mol^{-1} s^{-1}	k (400 K)/ l mol^{-1} s^{-1}
	$2,6-[(CH_3)_3]_2-4-CH_3O-C_6H_2O^{\bullet}$				
158	*cyclo*-$[-CH(NOH)(CH_2)_3-CH(CH_2)_3]$	−11.9	27.0	5.8×10^3	3.0×10^4
159	*cyclo*-$[-CH(NOH)(CH_2)_2-CH(CH_2)_3]$	−8.5	28.6	3.3×10^3	1.8×10^4
172	*cyclo*-$[C(CH_3)_2C(CH_3)_2N{=}C(C_6H_5)N(OH)]$	−18.5	23.9	1.8×10^4	7.6×10^4
173	*cyclo*-$[C(CH_3)_2C(C_6H_5){=}N(O)C(CH_3)[N(CH_3)OH]N(OH)]$	−20.7	22.9	2.6×10^4	1.0×10^5
200	$1,2-C_6H_3[C(C_6H_5)_3]$-*cyclo*-$[C(CH_3){=}CHC(CH_3)_2N(OH)]$	−41.2	14.2	5.9×10^5	1.4×10^6
201	$C_8H_{13}[(CH_3)_3C]NOH$	4.0	34.8	3.5×10^2	2.8×10^3
202	$C_6H_5C(O)[(CH_3)_3C]NOH$	−7.7	29.0	2.8×10^3	1.6×10^4
207	$[(CH_3)_3C]_2C{=}NOH$	7.8	36.7	1.8×10^2	1.6×10^3
208	$(CF_3)_2NOH$	15.0	40.5	44	5.1×10^2
209	$(CH_3)_2CHN(OH)C(CH_3)_3$	22.0	44.3	11	1.6×10^2
211	$(CH_3)_2CHCH_2C(O)N(OH)C(CH_3)_3$	−14.2	25.9	8.7×10^3	4.1×10^4
213	$C_6H_5CH{=}CHC(O)N(OH)C(CH_3)_3$	−9.3	28.2	3.8×10^3	2.1×10^4
214	$4-C_6H_5C_6H_4C(O)N(OH)C(CH_3)_3$	−6.3	29.6	2.3×10^3	1.4×10^4
215	$C_6H_5CH_2CH_2C(O)N(OH)C(CH_3)_3$	−12.1	26.9	6.0×10^3	3.1×10^4
217	*cyclo*-$[C(CH_3)_2(CH_2)_3C(CH_3)_2N(OH)]$	−39.0	15.1	4.3×10^5	1.1×10^6
218	*cyclo*-$[C(CH_3)_2CH_2CH(OCOC_6H_5)CH_2C(CH_3)_2N(OH)]$	−30.4	18.6	1.2×10^5	3.7×10^5
220	*cyclo*-$[C(CH_3)_2CHClC(O)CH_2C(CH_3)_2N(OH)]$	−22.8	22.0	3.5×10^4	1.3×10^5
221	*cyclo*-$[C(CH_3)_2CH_2CH(OH)CH_2C(CH_3)_2N(OH)]$	−28.5	19.5	8.7×10^4	2.8×10^5
222	*cyclo*-$[C(CH_3)_2CH_2C(O)CH_2C(CH_3)_2N(OH)]$	−30.3	18.7	1.2×10^5	3.6×10^5
227	*cyclo*-$[C(CH_3)_2C(O)N(OH)C(CH_3)_2N(OH)]$	−29.8	18.9	1.1×10^5	3.4×10^5
232	*cyclo*-$[C(CH_3)_2N(O){=}C(CH_3)CH_2C(CH_3)_2N(OH)]$	−21.9	22.4	3.1×10^4	1.2×10^5
235	*cyclo*-$[C(CH_3)_2CH_2C(C_6H_5){=}CHC(CH_3)_2N(OH)]$	−34.8	16.8	2.3×10^5	6.4×10^5
237	$1,2-C_6H_3[C(C_6H_5)_3]$-*cyclo*-$[C(CH_3)(C_6H_5)CH_2C(CH_3)_2N(OH)]$	−37.9	15.5	3.7×10^5	9.5×10^5

No.	Hydroxylamine	ΔH/ kJ mol^{-1}	E/ kJ mol^{-1}	k (333 K)/ l mol^{-1} s^{-1}	k (400 K)/ l mol^{-1} s^{-1}
	$2,6\text{-}[(CH_3)_3C]_2\text{-}4\text{-}CH_3\text{-}C_6H_2O^\bullet$				
158	*cyclo*-$[CH(NOH)(CH_2)_3\text{-}CH(CH_2)_3]$	−20.2	23.1	2.4×10^4	9.6×10^4
159	*cyclo*-$[CH(NOH)(CH_2)_2.CH(CH_2)_3]$	−16.8	24.7	1.3×10^4	6.0×10^4
172	*cyclo*-$[C(CH_3)_2C(CH_3)_2N{=}C(C_6H_5)N(OH)]$	−26.8	20.2	6.8×10^4	2.3×10^5
173	*cyclo*-$[C(CH_3)_2C(C_6H_5){=}N(O)C(CH_3)(N(CH_3)OH)N(OH)]$	−29.0	19.2	9.7×10^4	3.1×10^5
200	$1,2\text{-}C_6H_3[C(C_6H_5)_3]$-*cyclo*-$[C(CH_3){=}CHC(CH_3)_2N(OH)]$	−49.5	10.9	2.0×10^6	3.8×10^6
201	$C_8H_{13}[(CH_3)_3C]NOH$	−4.3	30.6	1.6×10^3	1.0×10^4
202	$C_6H_5C(O)[(CH_3)_3C]NOH$	−16.0	25.1	1.2×10^4	5.3×10^4
207	$[(CH_3)_3C]_2C{=}NOH$	−0.5	32.5	8.0×10^2	5.7×10^3
208	$(CF_3)_2NOH$	6.7	36.2	2.1×10^2	1.9×10^3
209	$(CH_3)_2CHN(OH)C(CH_3)_3$	13.7	39.8	57	6.3×10^2
211	$(CH_3)_2CHCH_2C(O)N(OH)C(CH_3)_3$	−22.5	22.1	3.4×10^4	1.3×10^5
213	$C_6H_5CH{=}CHC(O)N(OH)C(CH_3)_3$	−17.6	24.3	1.5×10^4	6.7×10^4
214	$4\text{-}C_6H_5C_6H_4C(O)N(OH)C(CH_3)_3$	−14.6	25.7	9.3×10^3	4.4×10^4
215	$C_6H_5CH_2CH_2C(O)N(OH)C(CH_3)_3$	−20.4	23.0	2.5×10^4	9.9×10^4
217	*cyclo*-$[C(CH_3)_2(CH_2)_3C(CH_3)_2N(OH)]$	−47.3	11.7	1.5×10^6	3.0×10^6
218	*cyclo*-$[C(CH_3)_2CH_2CH(OCOC_6H_5)CH_2C(CH_3)_2N(OH)]$	−38.7	15.2	4.1×10^5	1.0×10^6
220	*cyclo*-$[C(CH_3)_2CHClC(O)CH_2C(CH_3)_2N(OH)]$	−31.1	18.3	1.3×10^5	4.1×10^5
221	*cyclo*-$[C(CH_3)_2CH_2CH(OH)CH_2C(CH_3)_2N(OH)]$	−36.8	16.0	3.1×10^5	8.1×10^5
222	*cyclo*-$[C(CH_3)_2CH_2C(O)CH_2C(CH_3)_2N(OH)]$	−38.6	15.2	4.1×10^5	1.0×10^6
227	*cyclo*-$[C(CH_3)_2C(O)N(OH)C(CH_3)_2N(OH)]$	−38.1	15.4	3.8×10^5	9.7×10^5
232	*cyclo*-$[C(CH_3)_2N(O){=}C(CH_3)CH_2C(CH_3)_2N(OH)]$	−30.2	18.7	1.2×10^5	3.6×10^5
235	*cyclo*-$[C(CH_3)_2CH_2C(C_6H_5){=}CHC(CH_3)_2N(OH)]$	−43.1	13.4	7.9×10^5	1.8×10^6
237	$1,2\text{-}C_6H_3[C(C_6H_5)_3]$-*cyclo*-$[C(CH_3)(C_6H_5)CH_2C(CH_3)_2N(OH)]$	−46.2	12.2	1.2×10^6	2.6×10^6

No.	Hydroxylamine	ΔH/ kJ mol^{-1}	E/ kJ mol^{-1}	k (333 K)/ l mol^{-1} s^{-1}	k (400 K)/ l mol^{-1} s^{-1}
	2,6-[$(CH_3)_3C$]$_2$-4-CHO-$C_6H_2O^•$				
158	*cyclo*-[$CH(NOH)(CH_2)_3$-$CH(CH_2)_3$]	−29.0	19.2	9.7×10^4	3.1×10^5
159	*cyclo*-[$CH(NOH)(CH_2)_2CH(CH_2)_3$]	−25.6	20.7	5.7×10^4	2.0×10^5
172	*cyclo*-[$C(CH_3)_2C(CH_3)_2N{=}C(C_6H_5)N(OH)$]	−35.6	16.5	2.6×10^5	7.0×10^5
173	*cyclo*-[$C(CH_3)_2C(C_6H_5){=}N(O)C(CH_3)(N(CH_3)OH)N(OH)$]	−37.8	15.5	3.7×10^5	9.5×10^5
200	1,2-C_6H_3[$C(C_6H_5)_3$]-*cyclo*-[$C(CH_3){=}CHC(CH_3)_2N(OH)$]	−58.3	7.6	6.4×10^6	1.0×10^7
201	C_8H_{13}[$(CH_3)_3C$]NOH	−13.1	26.4	7.2×10^3	3.6×10^4
202	$C_6H_5C(O)$[$(CH_3)_3C$]NOH	−24.8	21.1	4.9×10^4	1.8×10^5
207	[$(CH_3)_3C$]$_2C{=}NOH$	−9.3	28.2	3.8×10^3	2.1×10^4
208	$(CF_3)_2NOH$	−2.1	31.7	1.1×10^3	7.2×10^3
209	$(CH_3)_2CHN(OH)C(CH_3)_3$	4.9	35.2	3.0×10^2	2.5×10^3
211	$(CH_3)_2CHCH_2C(O)N(OH)C(CH_3)_3$	−31.3	18.3	1.3×10^5	4.1×10^5
213	$C_6H_5CH{=}CHC(O)N(OH)C(CH_3)_3$	−26.4	20.4	6.3×10^4	2.2×10^5
214	4-$C_6H_5C_6H_4C(O)N(OH)C(CH_3)_3$	−23.4	21.7	3.9×10^4	1.5×10^5
215	$C_6H_5CH_2CH_2C(O)N(OH)C(CH_3)_3$	−29.2	19.2	9.7×10^4	3.1×10^5
217	*cyclo*-[$C(CH_3)_2(CH_2)_3C(CH_3)_2N(OH)$]	−56.1	8.4	4.8×10^6	8.0×10^6
218	*cyclo*-[$C(CH_3)_2CH_2CH(OCOC_6H_5)CH_2C(CH_3)_2N(OH)$]	−47.5	11.7	1.5×10^6	3.0×10^6
220	*cyclo*-[$C(CH_3)_2CHClC(O)CH_2C(CH_3)_2N(OH)$]	−39.9	14.7	4.9×10^5	1.2×10^6
221	*cyclo*-[$C(CH_3)_2CH_2CH(OH)CH_2C(CH_3)_2N(OH)$]	−45.6	12.4	1.1×10^6	2.4×10^6
222	*cyclo*-[$C(CH_3)_2CH_2C(O)CH_2C(CH_3)_2N(OH)$]	−47.4	11.7	1.5×10^6	3.0×10^6
227	*cyclo*-[$C(CH_3)_2C(O)N(OH)C(CH_3)_2N(OH)$]	−46.9	11.9	1.4×10^6	2.8×10^6
232	*cyclo*-[$C(CH_3)_2N(O){=}C(CH_3)CH_2C(CH_3)_2N(OH)$]	−39.0	15.1	4.3×10^5	1.1×10^6
235	*cyclo*-[$C(CH_3)_2CH_2C(C_6H_5){=}CHC(CH_3)_2N(OH)$]	−51.9	10.0	2.7×10^6	4.9×10^6
237	1,2-C_6H_3[$C(C_6H_5)_3$]-*cyclo*-[$C(CH_3)(C_6H_5)CH_2C(CH_3)_2N(OH)$]	−55.0	8.8	4.2×10^6	7.1×10^6

Table 5.7
Enthalpies, activation energies and rate constants of reactions of aminyl radical $(4\text{-}CH_3OC_6H_4)_2N^{\bullet}$ with hydroxylamines in nonpolar solutions calculated by formulas 1.15–1.17 and 1.20. The values of A, br_e and α, see Table 1.6

No.	Hydroxylamine	ΔH/ kJ mol^{-1}	E/ kJ mol^{-1}	k (333 K)/ l mol^{-1} s^{-1}	k (400 K)/ l mol^{-1} s^{-1}
158	*cyclo*-[-$CH(NOH)(CH_2)_3$-$CH(CH_2)_3$]	−22.4	0.8	7.5×10^6	7.9×10^6
159	*cyclo*-[-$CH(NOH)(CH_2)_2$-$CH(CH_2)_3$]	−19.0	4.8	1.8×10^6	2.4×10^6
172	*cyclo*-[$C(CH_3)_2C(CH_3)_2N{=}C(C_6H_5)N(OH)$]	−29.0	0.4	8.7×10^6	8.9×10^6
173	*cyclo*-[$C(CH_3)_2C(C_6H_5){=}N(O)C(CH_3)(N(CH_3)OH)N(OH)$]	−31.2	0.0	1.0×10^7	1.0×10^7
200	1,2-$C_6H_3[C(C_6H_5)_3]$-*cyclo*-[$C(CH_3){=}CHC(CH_3)_2N(OH)$]	−51.7	0.0	1.0×10^7	1.0×10^7
201	$C_8H_{13}[(CH_3)_3C]NOH$	−6.5	10.7	2.1×10^5	4.0×10^5
202	$C_6H_5C(O)[(CH_3)_3C]NOH$	−18.2	5.1	1.6×10^6	2.2×10^6
207	$[(CH_3)_3C]_2C{=}NOH$	−2.7	12.7	1.0×10^5	2.2×10^5
208	$(CF_3)_2NOH$	−4.5	11.7	1.5×10^5	3.0×10^5
209	$(CH_3)_2CHN(OH)C(CH_3)_3$	11.5	20.4	6.3×10^3	2.2×10^4
211	$(CH_3)_2CHCH_2C(O)N(OH)C(CH_3)_3$	−24.7	2.2	4.5×10^6	5.2×10^6
213	$C_6H_5CH{=}CHC(O)N(OH)C(CH_3)_3$	−19.8	4.3	2.1×10^6	2.7×10^6
214	4-$C_6H_5C_6H_4C(O)N(OH)C(CH_3)_3$	−16.8	5.8	1.2×10^6	1.7×10^6
215	$C_6H_5CH_2CH_2C(O)N(OH)C(CH_3)_3$	−22.6	3.1	3.3×10^6	3.9×10^6
221	*cyclo*-[$C(CH_3)_2CH_2CH(OH)CH_2C(CH_3)_2N(OH)$]	−39.0	0.0	1.0×10^7	1.0×10^7
222	*cyclo*-[$C(CH_3)_2CH_2C(O)CH_2C(CH_3)_2N(OH)$]	−40.8	0.0	1.0×10^7	1.0×10^7
227	*cyclo*-[$C(CH_3)_2C(O)N(OH)C(CH_3)_2N(OH)$]	−40.3	0.0	1.0×10^7	1.0×10^7
232	*cyclo*-[$C(CH_3)_2N(O){=}C(CH_3)CH_2C(CH_3)_2N(OH)$]	−32.4	0.0	1.0×10^7	1.0×10^7
235	*cyclo*-[$C(CH_3)_2CH_2C(C_6H_5){=}CHC(CH_3)_2N(OH)$]	−45.3	0.0	1.0×10^7	1.0×10^7
237	1,2-$C_6H_3[C(C_6H_5)_3]$- *cyclo*-[$C(CH_3)(C_6H_5)CH_2C(CH_3)_2N(OH)$]	−48.4	0.0	1.0×10^7	1.0×10^7

Table 5.8
Rate constants of reactions of alkyl radicals with nitroxyl radicals

Alkyl radical	Nitroxyl radical	*T*/ K	*k*/ l mol^{-1} s^{-1} or log *k* = *A* - *E*/θ	Ref.
$CH_3^•$	*cyclo*-[N(O$^•$)C(CH$_3$)$_2$CH$_2$CH(CONH$_2$)C(CH$_3$)$_2$]	298	5.1×10^7	8
$CH_3^•$	*cyclo*-[N(O$^•$)C(CH$_3$)$_2$CH=C(CONH$_2$)C(CH$_3$)$_2$]	298	7.8×10^8	8
$CH_3^•$	*cyclo*-[N(O$^•$)C(CH$_3$)$_2$CH(C(O)NH$_2$)C(CH$_3$)$_2$]	298	7.5×10^8	8
$CH_3^•$	*cyclo*-[N(O$^•$)C(CH$_3$)$_2$CH=C(C(O)NH$_2$)C(CH$_3$)$_2$]	298	7.8×10^8	8
CH$_3$(CH$_2$)$_7$C$^•$H$_2$	*cyclo*-[N(O$^•$)C(CH$_3$)$_2$(CH$_3$)$_3$C(CH$_3$)$_2$]	270–317	10.4 – 7.5/θ	8
CH$_3$(CH$_2$)$_7$C$^•$H$_2$	*cyclo*-[N(O$^•$)C(CH$_3$)$_2$(CH$_3$)$_3$C(CH$_3$)$_2$]	293	1.2×10^9	9
(CH$_3$)$_3$C$^•$	*cyclo*-[N(O$^•$)C(CH$_3$)$_2$(CH$_2$)$_3$]	293	7.6×10^8	9
(CH$_3$)$_3$C$^•$	1,2-*cyclo*-[C(CH$_3$)$_2$N(O$^•$)C(CH$_3$)$_2$]-C$_6$H$_4$	293	8.8×10^8	9
(CH$_3$)$_3$CC$^•$H$_2$	*cyclo*-[N(O$^•$)C(CH$_3$)$_2$(CH$_2$)$_3$C(CH$_3$)$_2$]	293	9.6×10^8	9
cyclo-[C$^•$H(CH$_2$)$_4$]	*cyclo*-[N(O$^•$)C(CH$_3$)$_2$CH$_2$CH(C(O)NH$_2$)C(CH$_3$)$_2$]	298	3.5×10^8	8
cyclo-[C$^•$H(CH$_2$)$_4$]	*cyclo*-[N(O$^•$)C(CH$_3$)$_2$CH=C(C(O)NH$_2$)C(CH$_3$)$_2$]	298	3.6×10^8	8
cyclo-[CH(C$^•$H$_2$)C(CH$_3$)$_2$(CH$_2$)$_2$]	1,2-*cyclo*-[C(CH$_3$)$_2$N(O$^•$)C(CH$_3$)$_2$-C$_6$H$_4$	333–398	10.0 – 2.5/θ	10
cyclo-[CH(C$^•$H$_2$)(CH$_2$)$_2$]	1,2-*cyclo*-[C(CH$_3$)$_2$N(O$^•$)C(CH$_3$)$_2$]-C$_6$H$_4$	333–398	7.4 – 4.6/θ	10
CH$_2$=CH(CH$_2$)$_3$C$^•$H$_2$	1,2-*cyclo*-[C(CH$_3$)$_2$N(O$^•$)C(CH$_3$)$_2$]-C$_6$H$_4$	333–398	9.7 – 3.8/θ	10
CH$_2$=CHC(CH$_3$)$_2$C$^•$H$_2$	1,2-*cyclo*-[C(CH$_3$)$_2$N(O$^•$)C(CH$_3$)$_2$]-C$_6$H$_4$	353	8.9 – 0.4/θ	10
CH$_2$=CHCH$_2$C(CH$_3$)$_2$CH$_2$C$^•$H$_2$	1,2-*cyclo*-[C(CH$_3$)$_2$N(O$^•$)C(CH$_3$)$_2$]-C$_6$H$_4$	333–398	10.0 – 4.2/θ	10
CH$_2$=CHCH$_2$OCH$_2$C$^•$(CH$_3$)$_2$	1,2-*cyclo*-[C(CH$_3$)$_2$N(O$^•$)C(CH$_3$)$_2$-C$_6$H$_4$	353	9.2 – 2.9/θ	10
CH$_2$=CHCH$_2$OC$^•$H$_2$	1,2-*cyclo*-[C(CH$_3$)$_2$N(O$^•$)C(CH$_3$)$_2$-C$_6$H$_4$	353	9.5 – 1.3/θ	10
CH$_2$=CHCH$_2$OCH$_2$C$^•$H$_2$	1,2-*cyclo*-[C(CH$_3$)$_2$N(O$^•$)C(CH$_3$)$_2$]-C$_6$H$_4$	333–398	9.6 – 2.1/θ	10
CH$_2$=CHCH$_2$OCH$_2$C$^•$HCH$_3$	1,2-*cyclo*-[C(CH$_3$)$_2$N(O$^•$)C(CH$_3$)$_2$]-C$_6$H$_4$	353	8.5 – 2.1/θ	10
C$_6$H$_5$C$^•$H$_2$	*cyclo*-[N(O$^•$)C(CH$_3$)$_2$(CH$_2$)$_3$C(CH$_3$)$_2$]	293	4.9×10^8	9
C$_6$H$_5$C$^•$H$_2$	*cyclo*-[N(O$^•$)C(CH$_3$)$_2$(CH$_2$)$_3$C(CH$_3$)$_2$]	233–306	9.3 – 3.7/θ	9
C$_6$H$_5$C$^•$H$_2$	1,2-*cyclo*-[C(CH$_3$)$_2$N(O$^•$)C(CH$_3$)$_2$]-C$_6$H$_4$	293	5.5×10^8	9
C$_6$H$_5$C$^•$HCH$_3$	*cyclo*-[N(O$^•$)C(CH$_3$)$_2$(CH$_2$)$_3$C(CH$_3$)$_2$]	293	1.6×10^8	9
C$_6$H$_5$C$^•$(CH$_3$)$_2$	*cyclo*-[N(O$^•$)C(CH$_3$)$_2$(CH$_2$)$_3$C(CH$_3$)$_2$]	293	1.2×10^9	9
(C$_6$H$_5$)$_2$C$^•$H	*cyclo*-[N(O$^•$)C(CH$_3$)$_2$(CH$_2$)$_3$C(CH$_3$)$_2$]	293	4.6×10^7	9

Alkyl radical	Nitroxyl radical	T/ K	k/ l mol⁻¹ s⁻¹ or log k = A - E/θ	Ref.
$(C_6H_5)_2C^{\bullet}CH_3$	*cyclo*-$[N(O^{\bullet})C(CH_3)_2(CH_2)_3C(CH_3)_2]$	293	4.6×10^7	9
$1\text{-}C^{\bullet}H_2\text{-}C_{10}H_7$	*cyclo*-$[N(O^{\bullet})C(CH_3)_2(CH_2)_3C(CH_3)_2]$	293	8.2×10^7	9
$2\text{-}C^{\bullet}H_2\text{-}C_{10}H_7$	*cyclo*-$[N(O^{\bullet})C(CH_3)_2(CH_2)_3C(CH_3)_2]$	293	5.7×10^7	9
$C^{\bullet}H_2OH$	*cyclo*-$[N(O^{\bullet})C(CH_3)_2CH_2CH(C(O)NH_2)C(CH_3)_2]$	298	4.6×10^8	8
$C^{\bullet}H_2OH$	*cyclo*-$[N(O^{\bullet})C(CH_3)_2CH{=}C(C(O)NH_2)C(CH_3)_2]$	298	3.5×10^8	8
$C^{\bullet}H_2OH$	*cyclo*-$[N(O^{\bullet})C(CH_3)_2CH{=}C(C(O)NH_2)C(CH_3)_2]$	298	3.5×10^8	8
$C^{\bullet}H_2OH$	*cyclo*-$[N(O^{\bullet})C(CH_3)_2CH{=}CH(C(O)NH_2)C(CH_3)_2]$	298	4.6×10^8	8
$C^{\bullet}H_2CH_2OH$	*cyclo*-$[N(O^{\bullet})C(CH_3)_2CH{=}CH(C(O)NH_2)C(CH_3)_2]$	298	4.7×10^8	8
$C^{\bullet}H_2CH_2OH$	*cyclo*-$[N(O^{\bullet})C(CH_3)_2CH{=}C(C(O)NH_2)C(CH_3)_2]$	298	4.8×10^8	8
$CH_3C^{\bullet}HOH$	*cyclo*-$[N(O^{\bullet})C(CH_3)_2CH{=}C(C(O)NH_2)C(CH_3)_2]$	298	6.2×10^8	8
$CH_3C^{\bullet}HOH$	*cyclo*-$[N(O^{\bullet})C(CH_3)_2CH_2CH(C(O)NH_2)C(CH_3)_2]$	298	4.3×10^8	8
$CH_3C^{\bullet}HOH$	*cyclo*-$[N(O^{\bullet})C(CH_3)_2CH_2CH(C(O)NH_2)C(CH_3)_2]$	298	4.7×10^8	8
$C^{\bullet}H_2C(CH_3)_2OH$	*cyclo*-$[N(O^{\bullet})C(CH_3)_2CH_2CH(C(O)NH_2)C(CH_3)_2]$	298	1.8×10^8	8
$C^{\bullet}H_2C(CH_3)_2OH$	*cyclo*-$[N(O^{\bullet})C(CH_3)_2CH{=}C(C(O)NH_2)C(CH_3)_2]$	298	2.0×10^8	8
$(CH_3)_2C^{\bullet}OH$	*cyclo*-$[N(O^{\bullet})C(CH_3)_2CH_2CH(C(O)NH_2)C(CH_3)_2]$	298	3.3×10^8	8
$(CH_3)_2C^{\bullet}OH$	*cyclo*-$[N(O^{\bullet})C(CH_3)_2CH{=}C(C(O)NH_2)C(CH_3)_2]$	298	3.6×10^8	8
$(CH_3)_2C^{\bullet}OH$	*cyclo*-$[N(O^{\bullet})C(CH_3)_2CH{=}C(C(O)NH_2)C(CH_3)_2]$	298	3.6×10^8	8
$(CH_3)_2C^{\bullet}OH$	*cyclo*-$[N(O^{\bullet})C(CH_3)_2CH_2CH(C(O)NH_2)C(CH_3)_2]$	298	3.3×10^8	8
$(C_6H_5)_2C^{\bullet}OH$	1,2-*cyclo*-[/C=C/NH- *cyclo*-$[C(CH_3)_2N(O^{\bullet})C(CH_3)_2CH_2]\text{-}C_{10}H_6$	295.5	2.6×10^7	11
$(C_6H_5)_2C^{\bullet}OH$	*cyclo*-$[N(O^{\bullet})C(CH_3)_2CBr{=}C(CN)C(CH_3)_2]$	295.5	5.0×10^7	11
$C^{\bullet}H_2C(CH_3)_2OH$	*cyclo*-$[N(O^{\bullet})C(CH_3)_2CH_2CH(C(O)NH_2)C(CH_3)_2]$	298	1.8×10^8	8
$CH_2(OH)(CHOH)_4C^{\bullet}(O)$	*cyclo*-$[N(O^{\bullet})C(CH_3)_2CH_2CH(C(OH)NH_2)C(CH_3)_2]$	298	5.1×10^7	8
$C^{\bullet}H_2C(CH_3)_2OH$	*cyclo*-$[N(O^{\bullet})C(CH_3)_2CH{=}C(C(OH)NH_2)C(CH_3)_2]$	298	2.0×10^8	8
$CH_2(OH)(CHOH)_4C^{\bullet}(O)$	*cyclo*-$[N(O^{\bullet})C(CH_3)_2CH{=}C(C(O)NH_2)C(CH_3)_2]$	298	4.3×10^7	8
$CH_2(OH)(CHOH)_4C^{\bullet}(O)$	*cyclo*-$[N(O^{\bullet})C(CH_3)_2CH_2CH(C(O)NH_2)C(CH_3)_2]$	298	5.1×10^7	8
$CH_2(OH)(CHOH)_4C^{\bullet}(O)$	*cyclo*-$[N(O^{\bullet})C(CH_3)_2CH{=}C(C(O)NH_2)C(CH_3)_2]$	298	4.3×10^7	8
cyclo-$[(CH_2)_3CH(C(CH_3)_2OC^{\bullet}(O)]$	1,2-*cyclo*-$[C(CH_3)_2N(O^{\bullet})C(CH_3)_2]\text{-}C_6H_4$	353	$9.7 - 2.5/\theta$	10

Solvent — H_2O,[8] Isooctane,[9] *cyclo*-C_6H_{12},[10] 1-Propanol.[11]

Table 5.9
Rate constants of disproportionation of nitroxyl radicals

Nitroxyl radical	Solvent	T/ K	$2k$/ $l\ mol^{-1}\ s^{-1}$ or $\log(2A) - E/\theta$	Ref.
$(CH_3)_2NO^{\bullet}$	CF_2Cl_2	158–298	$5.1 - 2.5/\theta$	12
$CH_3N(O^{\bullet})H$	C_6H_6	298	3.6×10^7	13
$(CH_3CH_2)_2NO^{\bullet}$	$(CH_3)_2CHCH_2CH_3$	298	6.6×10^4	14
$(CH_3CH_2)_2NO^{\bullet}$	C_6H_6	298	2.7×10^4	14
$(CH_3CH_2)_2NO^{\bullet}$	CF_2Cl_2	298	3.1×10^4	14
$(CH_3CH_2)_2NO^{\bullet}$	CH_3OH	298	3.0×10^3	14
$(CH_3CH_2)_2NO^{\bullet}$	H_2O	298	1.6×10^3	14
$(CH_3CH_2)_2NO^{\bullet}$	C_6H_6	298–348	$5.5 - 5.9/\theta$	14
$(CH_3CD_2)_2NO^{\bullet}$	$(CH_3)_2CHCH_2CH_3$	159–298	$4.2 - 2.5/\theta$	12
$CH_3CH_2N(O^{\bullet})CH(OH)CH_3$	H_2O	293	2.1×10^3	15
$[(CH_3)_2CH]_2NO^{\bullet}$	CF_2Cl_2	251–298	$7.9 - 39.3/\theta$	12
$(CH_3)_2CHN(O^{\bullet})H$	C_6H_6	301–341	$8.9 - 10.5/\theta$	13
$CH_3(CH_2)_5C(O)N(O^{\bullet})CH_3$	$C_6H_5CH_3$	—	$4.9 - 5.9/\theta$	16
$CH_3CH_2CH(CH_3)C(O)N(O^{\bullet})CH_3$	$CH_3(CH_2)_5CH_3$		$3.9 - 2.9/\theta$	16
$CH_3CH_2CH(CH_3)C(O)N(O^{\bullet})CD_3$	$CH_3(CH_2)_5CH_3$		$4.0 - 10.5/\theta$	16
$(CH_3)_3CH(O^{\bullet})H$	C_6H_6	287–345	$8.7 - 10.5/\theta$	13
$(CH_3)_3CC(O)N(O^{\bullet})CH_3$	$C_6H_5CH_3$		$4.3 - 13.0/\theta$	16
$(CH_3)_3C(CH_3)C{=}NO^{\bullet}$	C_6H_6	298	4.0×10^3	17
$[(CH_3)_3C]_2C{=}NO^{\bullet}$	C_6H_6	298–328	$5.3 - 20.9/\theta$	17
$(CH_3CH_2)_3CC(O)N(O^{\bullet})CH_3$	$C_6H_5CH_3$		$4.6 - 15.1/\theta$	16
$CH_3(CH_2)_5N(O^{\bullet})C(CH_3)_3$	C_6H_6	313	1.2×10^2	18
$CH_3(CH_2)_5(C_6H_5)CHN(O^{\bullet})C(CH_3)_3$	C_6H_6	313	1.0	18
$[(CH_3)_3C]_2C{=}NO^{\bullet}$	C_6H_6	297	2.1×10^5	19
$2,3,5,6\text{-}(CH_3)_4C_6HN(O^{\bullet})(CH_2)_5CH_3$	C_6H_6	313	50	18
$(C_6H_5)_2C{=}NO^{\bullet}$	C_6H_5Cl	288–333	$5.3 - 8.4/\theta$	17
$C_6H_5N(O^{\bullet})H$	C_6H_6	298	3.4×10^6	13
$C_6H_5(CH_3)C{=}NO^{\bullet}$	C_6H_6	298–313	$9.3 - 20.9/\theta$	17

Nitroxyl radical	Solvent	T/ K	$2k$/ l mol^{-1} s^{-1} or log(2A) −E/θ	Ref.
4-$CH_3C_6H_4(CH_3)C{=}NO^•$	C_6H_6	298	1.0×10^5	17
4-$CH_3OC_6H_4(CH_3)C{=}NO^•$	CH_2Cl_2	298	4.0×10^5	17
4-$NH_2C_6H_4(CH_3)C{=}NO^•$	C_6H_6	298	1.0×10^6	17
4-$NO_2C_6H_4(CH_3)C{=}NO^•$	C_6H_6	298	6.0×10^4	17
4-$BrC_6H_4(CH_3)C{=}NO^•$	C_6H_6	298	8.0×10^4	17
4-$CNC_6H_4(CH_3)C{=}NO^•$	C_6H_6	298	4.0×10^4	17
$C_6H_5CH{=}NO^•$	C_6H_6	298	1.8×10^8	17
$C_6H_5(CH_3CH_2)C{=}NO^•$	C_6H_6	298	6.0×10^4	17
$(CH_3)_2C(C_6H_5)C{=}NO^•$	C_6H_6	298	4.0×10^2	17
(4-$CH_3OC_6H_4)_2C{=}NO^•$	C_6H_6	298	6.0×10^4	17
(4-$CH_3C_6H_4)_2C{=}NO^•$	C_6H_6	298	1.0×10^4	17
(4-$ClC_6H_4)_2C{=}NO^•$	C_6H_6	298	8.0×10^3	17
2-$C_{10}H_7N(O^•)C(CH_3)_3$	CCl_4	318	1.1×10^{-4}	20
8-$CH_3C_{10}H_6$-2-$N(O^•)C(CH_3)_3$	CCl_4	303	2.1×10^{-5}	20
6-$CH_3OC_{10}H_6$-2-$N(O^•)C(CH_3)_3$	CCl_4	303	7.7×10^{-7}	20
4.5-$(CH_3)_2$-$C_{10}H_5$-2-$N(O^•)C(CH_3)_3$	CCl_4	303	2.5×10^{-4}	20
cyclo-[$(CH_2)_4N(O^•)$]	$(CH_3)_2CHCH_2CH_3$	298	1.8×10^5	12
cyclo-[$(CH_2)_5N(O^•)$]	$(CH_3)_2CHCH_2CH_3$	201–298	5.3 + 3.1/θ	12
cyclo-[$(CH_2)_3C(CH_3)_2NO^•$]	C_6H_6	298	2.4×10^3	21
cyclo-[$(CH_3)_2C(CH_2)_2CH(OC(CH_3)_3)N(O^•)$]	$[(CH_3)_3CO]_2$	298	5.1×10^2	22
cyclo-[$(CH_3)(CH_2CH_2CH_3)C(CH_2)_2CH(OC(CH_3)_3)N(O^•)$]	$[(CH_3)_3CO]_2$	298	3.9×10^2	22
cyclo-[$(CH_3)CH(CH_2)_2CH(OC(CH_3)_3)N(O^•)$]	$[(CH_3)_3CO]_2$	298	5.5×10^2	22
cyclo-[$CD(CH_3)(CH_3)_2CH(OC(CH_3)_3)N(O^•)$]	$[(CH_3)_3CO]_2$	298	4.1×10^2	22
cyclo-[$C(CH_3)_2CH_2C(CH_3)_2CH(OC(CH_3)_3)N(O^•)$]	$[(CH_3)_3CO]_2$	298	65	22

Table 5.10
Enthalpies, activation energies and rate constants of reactions of nitroxyl radicals with hydrocarbons and hydroperoxides in hydrocarbon solutions calculated by formulas 1.15–1.17, 1.20 and 1.21. The values of A, br_e and α, see Table 1.6

Hydrocarbon, hydroperoxide	ΔH/ kJ mol^{-1}	E/ kJ mol^{-1}	k (333 K)/ l mol^{-1} s^{-1}	k (400 K)/ l mol^{-1} s^{-1}
		$[(CH_3)_3C]_2C{=}NO^{\bullet}$		
cyclo-C_5H_{10}	56.9	68.6	0.18	0.51
cyclo-C_6H_{12}	61.1	71.0	8.7×10^{-2}	6.2
cyclo-$C_6H_{11}CH_3$	50.8	65.2	5.9×10^{-2}	9.2
cis-$CH_3CH{=}CHCH_3$	12.5	63.3	7.0×10^{-2}	3.2
$CH_2{=}C(CH_3)CH_2CH_3$	4.5	59.9	8.0×10^{-2}	3.0
cyclo-C_6H_{10}	−1.2	57.4	4.0	1.2×10^{2}
$(CH_3)_2C{=}C(CH_3)_2$	10.2	62.4	0.2	8.3
$CH_2{=}CHCH(CH_3)_2$	−3.1	56.5	0.14	4.1
1,3-Cyclohexadiene	−33.1	44.1	4.8×10^{2}	6.6×10^{3}
$C_6H_5CH_3$	29.7	60.1	4.0×10^{-2}	4.2
4-$CH_3C_6H_4CH_3$	26.9	58.7	0.37	12.7
$C_6H_5CH_2CH_3$	18.5	54.7	0.52	14.1
$C_6H_5CH(CH_3)_2$	12.1	51.7	0.78	17.4
$C_6H_5CH_2CH{=}CH_2$	4.1	48.0	5.8	1.0×10^{2}
$(C_6H_5)_2CH_2$	11.5	51.4	1.7	38.1
$(C_6H_5)_2CHCH_3$	5.7	48.9	2.1	40.4
cyclo-$C_6H_{11}C_6H_5$	11.1	51.3	0.9	19.7
Tetraline	10.5	50.9	40.4	9.2×10^{2}
Indane	16.3	53.5	15.8	4.1×10^{2}
1-CH_3-Tetralyl	−6.6	43.3	47.3	6.6×10^{2}
$(C_6H_5)_3CH$	4.3	48.1	2.8	51.5
9,10-Dihydroanthracene	−23.3	36.6	7.2×10^{3}	6.6×10^{4}
1,2-Dihydronaphthalene	−8.8	42.4	4.4×10^{2}	5.7×10^{3}
9,10-Dihydrofluorene	−1.6	45.5	2.8×10^{2}	4.5×10^{3}

Hydrocarbon, hydroperoxide	ΔH/ kJ mol^{-1}	E/ kJ mol^{-1}	k (333 K)/ l mol^{-1} s^{-1}	k (400 K)/ l mol^{-1} s^{-1}
H_2O_2	30.5	57.1	0.12	2.2
sec-ROOH	27.0	55.1	0.12	2.0
tert-ROOH	20.1	51.4	0.24	6.1
***cyclo*-[C(CH$_3$)$_2$CH$_2$C(O)CH$_2$C(CH$_3$)$_2$N(O$^•$)]**				
cyclo-C_5H_{10}	95.0	96.4	7.8×10^{-6}	2.6×10^{-3}
cyclo-C_6H_{12}	99.2	100.8	2.0×10^{-6}	8.5×10^{-4}
cyclo-$C_6H_{11}CH_3$	85.9	87.3	2.0×10^{-5}	4.8×10^{-3}
cis-$CH_3CH{=}CHCH_3$	50.6	82.0	8.4×10^{-5}	1.2×10^{-2}
$CH_2{=}C(CH_3)CH_2CH_3$	42.6	77.9	1.2×10^{-4}	1.3×10^{-2}
cyclo-C_6H_{10}	36.9	74.1	6.6×10^{-3}	0.62
$(CH_3)_2C{=}C(CH_3)_2$	48.3	80.9	2.5×10^{-4}	3.3×10^{-2}
$CH_2{=}CHCH(CH_3)_2$	35.0	75.7	1.3×10^{-4}	1.3×10^{-2}
1,3-Cyclohexadiene	5.0	60.0	1.6	58.4
$C_6H_5CH_3$	67.8	80.6	6.6×10^{-5}	9.2×10^{-3}
4-$CH_3C_6H_4CH_3$	65.0	78.9	2.5×10^{-4}	3.0×10^{-2}
$C_6H_5CH_2CH_3$	56.6	74.2	4.6×10^{-4}	4.1×10^{-2}
$C_6H_5CH(CH_3)_2$	50.2	70.7	7.8×10^{-4}	5.8×10^{-2}
$C_6H_5CH_2CH{=}CH_2$	42.2	66.5	7.2×10^{-3}	0.41
$(C_6H_5)_2CH_2$	49.6	70.4	1.8×10^{-3}	0.12
$(C_6H_5)_2CHCH_3$	43.8	67.3	5.5×10^{-3}	0.16
cyclo-$C_6H_{11}C_6H_5$	49.2	70.2	9.6×10^{-4}	6.6×10^{-2}
Tetraline	48.6	69.6	4.8×10^{-2}	3.3
Indane	54.4	73.0	1.4×10^{-2}	1.2
1-CH_3-Tetralyl	31.5	61.0	0.27	10.8
$(C_6H_5)_3CH$	42.4	66.6	3.5×10^{-3}	0.2
9,10-Dihydroanthracene	14.8	52.9	21.9	4.9×10^{-2}
1,2-Dihydronaphthalene	29.3	59.9	0.24	60.4

Hydrocarbon, hydroperoxide	**ΔH/ kJ mol^{-1}**	**E/ kJ mol^{-1}**	**k (333 K)/ l mol^{-1} s^{-1}**	**k (400 K)/ l mol^{-1} s^{-1}**
9,10-Dihydrofluorene	36.5	63.5	0.42	20.3
H_2O_2	68.6	80.0	1.8×10^{-5}	2.2×10^{-3}
sec-ROOH	65.1	77.8	2.0×10^{-5}	2.2×10^{-3}
tert-ROOH	58.2	73.5	9.0×10^{-5}	7.9×10^{-3}
***cyclo*-[C(CH$_3$)$_2$C(CH$_3$)$_2$N=C(C$_6$H$_5$)N(O$^•$)]**				
cyclo-C_5H_{10}	83.2	84.4	5.8×10^{-4}	9.5×10^{-2}
cyclo-C_6H_{12}	87.4	87.4	2.3×10^{-4}	4.6×10^{-2}
cyclo-$C_6H_{11}CH_3$	77.1	80.6	4.6×10^{-4}	7.6×10^{-2}
cis-$CH_3CH{=}CHCH_3$	38.1	75.6	8.3×10^{-4}	8.0×10^{-2}
$CH_2{=}C(CH_3)CH_2CH_3$	30.8	72.0	1.0×10^{-3}	7.9×10^{-2}
cyclo-C_6H_{10}	25.1	69.2	5.6×10^{-2}	3.7
$(CH_3)_2C{=}C(CH_3)_2$	36.5	74.8	2.2×10^{-3}	0.20
$CH_2{=}CHCH(CH_3)_2$	23.2	68.2	1.9×10^{-3}	0.12
1,3-Cyclohexadiene	−6.8	54.8	10.1	2.8×10^{2}
$C_6H_5CH_3$	56.0	73.9	7.7×10^{-4}	6.7×10^{-2}
4-$CH_3C_6H_4CH_3$	53.2	72.3	2.7×10^{-3}	0.22
$C_6H_5CH_2CH_3$	44.8	67.8	4.6×10^{-3}	0.28
$C_6H_5CH(CH_3)_2$	38.4	64.5	7.6×10^{-3}	0.38
$C_6H_5CH_2CH{=}CH_2$	30.4	60.4	6.7×10^{-2}	2.6
$(C_6H_5)_2CH_2$	37.8	64.2	1.7×10^{-2}	0.83
$(C_6H_5)_2CHCH_3$	32.0	61.2	2.5×10^{-2}	1.0
cyclo-$C_6H_{11}C_6H_5$	37.4	64.0	9.1×10^{-3}	0.44
Tetraline	36.8	63.7	0.41	19.2
Indane	42.6	66.7	0.14	7.8
1-CH_3-Tetralyl	19.7	55.2	2.2	62.0
$(C_6H_5)_3CH$	30.6	60.5	3.2×10^{-2}	1.3
9,10-Dihydroanthracene	3.0	47.6	1.4×10^{2}	2.4×10^{3}

Hydrocarbon, hydroperoxide	ΔH/ kJ mol^{-1}	E/ kJ mol^{-1}	k (333 K)/ l mol^{-1} s^{-1}	k (400 K)/ l mol^{-1} s^{-1}
1,2-Dihydronaphthalene	17.5	54.2	12.6	3.3×10^{2}
9,10-Dihydrofluorene	24.7	57.6	3.7	1.2×10^{2}
H_2O_2	56.8	72.6	2.6×10^{-4}	2.1×10^{-2}
sec-ROOH	53.3	70.5	2.8×10^{-4}	2.0×10^{-2}
tert-ROOH	46.4	66.3	1.3×10^{-3}	7.0×10^{-2}
***cyclo*-[C(CH$_3$)$_2$CH(CH$_3$)N(CH$_3$)C(CH$_3$)$_2$N(O$^\bullet$)]**				
cyclo-C_5H_{10}	104.6	104.6	3.9×10^{-7}	2.2×10^{-4}
cyclo-C_6H_{12}	108.8	108.8	1.0×10^{-7}	7.4×10^{-5}
cyclo-$C_6H_{11}CH_3$	98.5	98.5	4.5×10^{-7}	2.1×10^{-4}
cis-$CH_3CH{=}CHCH_3$	60.2	87.2	1.3×10^{-5}	2.5×10^{-3}
$CH_2{=}C(CH_3)CH_2CH_3$	52.2	82.9	2.0×10^{-5}	3.0×10^{-3}
cyclo-C_6H_{10}	46.5	80.0	1.1×10^{-3}	0.14
$(CH_3)_2C{=}C(CH_3)_2$	57.9	86.0	3.9×10^{-5}	7.0×10^{-3}
$CH_2{=}CHCH(CH_3)_2$	44.6	78.9	4.2×10^{-5}	5.0×10^{-3}
1,3-Cyclohexadiene	14.6	64.3	0.33	16.0
$C_6H_5CH_3$	77.4	86.2	9.0×10^{-6}	1.7×10^{-3}
4-$CH_3C_6H_4CH_3$	74.6	84.5	3.3×10^{-5}	5.5×10^{-3}
$C_6H_5CH_2CH_3$	66.2	79.6	6.5×10^{-5}	8.1×10^{-3}
$C_6H_5CH(CH_3)_2$	59.8	76.0	1.2×10^{-4}	1.2×10^{-2}
$C_6H_5CH_2CH{=}CH_2$	51.8	71.6	1.2×10^{-3}	8.9×10^{-2}
$(C_6H_5)_2CH_2$	59.2	75.7	2.7×10^{-4}	2.6×10^{-2}
$(C_6H_5)_2CHCH_3$	53.4	72.4	4.4×10^{-4}	3.5×10^{-2}
cyclo-$C_6H_{11}C_6H_5$	58.8	75.4	1.5×10^{-4}	1.4×10^{-2}
Tetraline	58.2	75.1	6.6×10^{-3}	0.62
Indane	64.0	78.4	2.0×10^{-3}	0.23
1-CH_3-Tetralyl	41.1	65.9	4.6×10^{-2}	2.5
$(C_6H_5)_3CH$	52.0	71.7	5.7×10^{-4}	4.3×10^{-2}

Hydrocarbon, hydroperoxide	ΔH/ kJ mol^{-1}	E/ kJ mol^{-1}	k (333 K)/ l mol^{-1} s^{-1}	k (400 K)/ l mol^{-1} s^{-1}
9,10-Dihydroanthracene	24.4	57.5	3.8	1.2×10^{2}
1,2-Dihydronaphthalene	38.9	64.7	0.3	14.2
9,10-Dihydrofluorene	46.1	68.5	7.2×10^{-2}	4.5
H_2O_2	78.2	86.3	1.9×10^{-6}	3.4×10^{-4}
sec-ROOH	74.7	84.0	2.1×10^{-6}	3.4×10^{-4}
tert-ROOH	67.8	79.5	1.1×10^{-5}	1.3×10^{-3}

Table 5.11
Enthalpies, activation energies and rate constants of reactions of nitroxyl radicals with phenols in hydrocarbon solutions calculated by formulas 1.15–1.17 and 1.21. The values of A, br_e and α, see Table 1.6

No.	Phenol	ΔH/ kJ mol^{-1}	E/ kJ mol^{-1}	k (333 K)/ l mol^{-1} s^{-1}	k (400 K)/ l mol^{-1} s^{-1}
	$[(CH_3)_3C]_2C{=}NO^{\bullet}$				
3	$2\text{-}HOC_6H_4OH$	−0.5	21.4	4.4×10^{4}	1.6×10^{5}
4	$2\text{-}HO\text{-}4\text{-}(CH_3)_3CC_6H_3OH$	3.9	23.7	1.9×10^{4}	8.0×10^{4}
9	$4\text{-}CH_3C_6H_4OH$	19.3	31.9	9.9×10^{2}	6.8×10^{3}
10	$(CH_3)_3C_6(OH)(CH_2)_3O$	−9.7	17.0	2.2×10^{5}	6.0×10^{5}
20	$(CH_3)_3C_6(OH)CH_2CH(CH_3)O$	−14.1	14.9	4.6×10^{5}	1.1×10^{6}
25	$1\text{-}HOC_{13}H_9$	−2.2	20.6	5.9×10^{4}	2.0×10^{5}
29	$1\text{-}HOC_{10}H_7$	2.9	23.2	2.3×10^{4}	9.3×10^{4}
30	$2\text{-}HOC_{10}H_7$	13.3	28.6	3.3×10^{3}	1.8×10^{4}
32	$1\text{-}HOC_{14}H_9$	14.2	29.1	2.7×10^{3}	1.6×10^{4}
36	C_6H_5OH	28.5	37.2	1.5×10^{2}	1.4×10^{3}
38	$4\text{-}CH_3COC_6H_4OH$	31.0	38.6	0.88	9.1×10^{2}
40	$4\text{-}NH_2C_6H_4OH$	16.2	30.2	1.8×10^{3}	1.1×10^{4}
41	$4\text{-}C_6H_5CH_2OC_6H_4OH$	9.0	26.3	7.5×10^{3}	3.7×10^{4}
44	$4\text{-}CH_3(CH_2)_3OC_6H_4OH$	7.3	25.4	1.0×10^{4}	4.8×10^{4}
46	$4\text{-}(CH_3)_3CC_6H_4OH$	18.4	31.4	1.2×10^{3}	7.9×10^{3}

No.	Phenol	ΔH/ kJ mol^{-1}	E/ kJ mol^{-1}	k (333 K)/ l mol^{-1} s^{-1}	k (400 K)/ l mol^{-1} s^{-1}
47	2,4-$[(CH_3)_3C]_2C_6H_3OH$	19.0	31.7	1.1×10^3	7.3×10^3
74	2-$(CH_3)_3C$-4,6-$(CH_3)_2C_6H_2OH$	15.3	29.7	2.2×10^3	1.3×10^4
75	2-$(CH_3)_3C$-4-$CH_3OC_6H_3OH$	2.9	23.2	2.3×10^4	9.3×10^4
78	4-$CH_3OC_6H_4OH$	10.5	27.1	5.6×10^3	2.9×10^4
79	2,3-$(CH_3)_2C_6H_3OH$	15.0	29.5	2.4×10^3	1.4×10^4
80	3,4-$(CH_3)_2C_6H_3OH$	14.0	29.0	2.8×10^3	1.6×10^4
81	3,5-$(CH_3)_2C_6H_3OH$	15.5	29.8	2.1×10^3	1.3×10^4
82	2,4-$(CH_3)_2C_6H_3OH$	20.0	32.3	8.6×10^2	6.1×10^3
83	2,6-$(CH_3)_2C_6H_3OH$	16.4	30.3	1.8×10^3	1.1×10^4
85	4-$(CH_3)_2NC_6H_4OH$	−11.7	16.1	3.0×10^5	7.9×10^5
110	2,4,6-$(CH_3)_3C_6H_2OH$	8.3	25.9	8.7×10^3	4.1×10^4
130	α-$(CH_3)_3C_6(OH)(CH_2)_2OC(CH_3)[(CH_2)_3CH(CH_3)]_3CH_3$	−10.8	16.5	2.6×10^5	7.0×10^5
131	β-$(CH_3)_2C_6H(OH)(CH_2)_2OC(CH_3)[(CH_2)_3CH(CH_3)]_3CH_3$	−6.1	18.7	1.2×10^5	3.6×10^5
133	δ-$CH_3C_6H_2(OH)(CH_2)_2OC(CH_3)[(CH_2)_3CH(CH_3)]_3CH_3$	−21.5	11.6	1.5×10^6	3.1×10^6
245	C_6H_5SH	10.0	33.1	6.4×10^2	4.8×10^3
	cyclo-$[C(CH_3)_2C(CH_3)_2N{=}C(C_6H_5)N(O^\bullet)]$				
3	2-HOC_6H_4OH	25.8	35.6	2.6×10^2	2.2×10^3
4	2-HO-4-$(CH_3)_3CC_6H_3OH$	30.2	38.2	1.0×10^2	1.0×10^3
9	4-$CH_3C_6H_4OH$	45.6	47.6	3.4	0.61
10	$(CH_3)_3C_6(OH)(CH_2)_3O$	16.6	30.4	1.7×10^3	1.1×10^4
20	$(CH_3)_3C_6(OH)CH_2CH(CH_3)O$	12.2	28.0	4.1×10^3	2.2×10^4
25	1-$HOC_{13}H_9$	24.1	34.6	3.7×10^2	3.0×10^3
29	1-$HOC_{10}H_7$	29.2	37.6	1.3×10^2	1.2×10^3
30	2-$HOC_{10}H_7$	39.6	43.8	13.5	1.9×10^2
32	1-$HOC_{14}H_9$	40.5	44.4	10.8	1.6×10^2
36	C_6H_5OH	54.8	54.8	0.25	7.0
38	4-$CH_3COC_6H_4OH$	57.3	57.3	0.10	3.3

No.	Phenol	ΔH/ kJ mol^{-1}	E/ kJ mol^{-1}	k (333 K)/ l mol^{-1} s^{-1}	k (400 K)/ l mol^{-1} s^{-1}
40	$4\text{-}NH_2C_6H_4OH$	42.5	45.6	7.0	1.1×10^2
41	$4\text{-}C_6H_5CH_2OC_6H_4OH$	35.6	41.2	0.34	4.2×10^2
44	$4\text{-}CH_3(CH_2)_3OC_6H_4OH$	33.6	40.2	0.49	5.6×10^2
46	$4\text{-}(CH_3)_3CC_6H_4OH$	44.7	47.0	4.2	73
47	$2,4\text{-}[(CH_3)_3C]_2C_6H_3OH$	45.3	47.4	3.7	65
74	$2\text{-}(CH_3)_3C\text{-}4,6\text{-}(CH_3)_2C_6H_2OH$	41.6	45.1	8.4	1.3×10^2
75	$2\text{-}(CH_3)_3C\text{-}4\text{-}CH_3OC_6H_3OH$	29.2	37.6	1.3×10^{-2}	1.2×10^3
78	$4\text{-}CH_3OC_6H_4OH$	36.8	42.1	0.25	3.2×10^2
79	$2,3\text{-}(CH_3)_2C_6H_3OH$	41.3	44.9	9.1	1.4×10^2
80	$3,4\text{-}(CH_3)_2C_6H_3OH$	40.4	44.3	0.11	1.6×10^2
81	$3,5\text{-}(CH_3)_2C_6H_3OH$	41.8	45.2	8.1	1.3×10^2
82	$2,4\text{-}(CH_3)_2C_6H_3OH$	46.3	48.1	2.8	52
83	$2,6\text{-}(CH_3)_2C_6H_3OH$	42.7	45.8	6.5	1.0×10^2
85	$4\text{-}(CH_3)_2NC_6H_4OH$	14.6	29.3	2.5×10^3	1.5×10^4
110	$2,4,6\text{-}(CH_3)_3C_6H_2OH$	34.6	40.8	40	4.7×10^2
130	$\alpha\text{-}(CH_3)_3C_6(OH)(CH_2)_2OC(CH_3)[(CH_2)_3CH(CH_3)]_3CH_3$	15.5	29.8	2.1×10^3	1.3×10^4
131	$\beta\text{-}(CH_3)_2C_6H(OH)(CH_2)_2OC(CH_3)[(CH_2)_3CH(CH_3)]_3CH_3$	20.2	32.4	8.3×10^2	5.9×10^4
133	$\delta\text{-}CH_3C_6H_2(OH)(CH_2)_2OC(CH_3)[(CH_2)_3CH(CH_3)]_3CH_3$	4.2	24.1	1.7×10^4	7.1×10^4
245	C_6H_5SH	36.3	45.4	7.56	1.2×10^2
	***cyclo*-[C(CH$_3$)$_2$CH$_2$C(O)CH$_2$C(CH$_3$)$_2$N(O$^\bullet$)]**				
3	$2\text{-}HOC_6H_4OH$	37.6	42.6	21.0	2.7×10^3
4	$2\text{-}HO\text{-}4\text{-}(CH_3)_3CC_6H_3OH$	42.0	45.3	7.8	1.2×10^2
9	$4\text{-}CH_3C_6H_4OH$	57.4	57.4	9.9×10^{-2}	3.2
10	$(CH_3)_3C_6(OH)(CH_2)_3O$	28.4	37.1	1.5×10^2	1.4×10^3
20	$(CH_3)_3C_6(OH)CH_2CH(CH_3)O$	24.0	34.6	3.7×10^2	3.0×10^3
25	$1\text{-}HOC_{13}H_9$	35.9	41.6	29.8	3.7×10^2
29	$1\text{-}HOC_{10}H_7$	41.0	44.7	9.7	1.5×10^2

No.	Phenol	ΔH/ kJ mol^{-1}	E/ kJ mol^{-1}	k (333 K)/ l mol^{-1} s^{-1}	k (400 K)/ l mol^{-1} s^{-1}
30	$2\text{-}HOC_{10}H_7$	51.4	51.4	0.87	0.19
32	$1\text{-}HOC_{14}H_9$	52.3	52.3	0.62	0.15
36	C_6H_5OH	66.6	66.6	3.6×10^{-3}	0.20
38	$4\text{-}CH_3COC_6H_4OH$	69.1	69.1	1.4×10^{-3}	9.5×10^{-2}
40	$4\text{-}NH_2C_6H_4OH$	54.3	54.3	0.30	8.1
41	$4\text{-}C_6H_5CH_2OC_6H_4OH$	47.1	48.6	2.4	45
44	$4\text{-}CH_3(CH_2)_3OC_6H_4OH$	45.4	47.5	3.5	63
46	$4\text{-}(CH_3)_3CC_6H_4OH$	56.5	56.5	0.14	4.2
47	$2,4\text{-}[(CH_3)_3C]_2C_6H_3OH$	57.1	57.1	0.11	3.5
74	$2\text{-}(CH_3)_3C\text{-}4,6\text{-}(CH_3)_2C_6H_2OH$	53.4	53.4	0.42	10.6
75	$2\text{-}(CH_3)_3C\text{-}4\text{-}CH_3OC_6H_3OH$	41.0	44.7	9.7	1.5×10^{2}
78	$4\text{-}CH_3OC_6H_4OH$	48.6	49.5	1.7	34.4
79	$2,3\text{-}(CH_3)_2C_6H_3OH$	53.1	53.1	0.47	11.6
80	$3,4\text{-}(CH_3)_2C_6H_3OH$	52.2	52.2	0.65	15.3
81	$3,5\text{-}(CH_3)_2C_6H_3OH$	53.6	53.6	0.39	10.0
82	$2,4\text{-}(CH_3)_2C_6H_3OH$	58.1	58.1	7.7×10^{-2}	2.6
83	$2,6\text{-}(CH_3)_2C_6H_3OH$	54.5	54.5	0.28	7.6
85	$4\text{-}(CH_3)_2NC_6H_4OH$	26.4	35.9	2.3×10^{2}	2.1×10^{3}
110	$2,4,6\text{-}(CH_3)_3C_6H_2OH$	46.4	48.1	2.85	52.3
130	$\alpha\text{-}(CH_3)_3C_6(OH)(CH_2)_2OC(CH_3)[(CH_2)_3CH(CH_3)]_3CH_3$	27.3	36.5	1.9×10^{2}	1.7×10^{3}
131	$\beta\text{-}(CH_3)_2C_6H(OH)(CH_2)_2OC(CH_3)[(CH_2)_3CH(CH_3)]_3CH_3$	32.0	39.2	71.0	7.6×10^{2}
133	$\delta\text{-}CH_3C_6H_2(OH)(CH_2)_2OC(CH_3)[(CH_2)_3CH(CH_3)]_3CH_3$	16.6	30.4	1.7×10^{3}	1.1×10^{4}
245	C_6H_5SH	48.1	51.6	0.81	18

Table 5.12
Enthalpies, activation energies and rate constants of reactions of nitroxyl radicals with sterically hindered phenols in hydrocarbon solutions calculated by formulas 1.15–1.17 and 1.21. The values of *A*, br_e and α, see Table 1.6

No.	Phenol	ΔH/ kJ mol^{-1}	E/ kJ mol^{-1}	k (333 K)/ l mol^{-1} s^{-1}	k (400 K)/ l mol^{-1} s^{-1}
	$[(CH_3)_3C]_2C{=}NO^{\bullet}$				
1	1,3,5-[3′,5′-$[(CH_3)_2C]_2$-4′-HO-$C_6H_3CH_2]_3C_6H_3$	6.4	36.0	2.3×10^3	2.3×10^3
5	2-HO-3,5-$[(CH_3)_3C]_2C_6H_2OH$	−0.2	32.7	7.4×10^2	5.4×10^3
24	2,2′-CH_2-[4,6-$[(CH_3)_3C]_2C_6H_2OH]_2$	2.4	34.0	4.6×10^2	3.6×10^3
31	C-[4-$(CH_2)_2COOCH_2$-2,6-$[(CH_3)_3C]_3C_6H_2OH]$	2.5	34.0	4.6×10^2	3.6×10^3
49	2,4-$[(CH_3)_3C]_2$-6-CH_3-C_6H_2OH	15.4	40.7	41.3	4.8×10^2
50	4,4′-[2,6-$[(CH_3)_3C]_2C_6H_2OH]_2$	4.1	34.8	3.5×10^2	2.9×10^3
51	2,6-$[(CH_3)_3C]_2C_6H_3OH$	8.2	36.9	1.6×10^2	1.5×10^3
52	2,6-$[(CH_3)_3C]_2$-4-$CH_3CONHC_6H_2OH$	−11.8	27.0	5.8×10^3	3.0×10^4
53	2,6-$[(CH_3)_3C]_2$-4-$CH_3COC_6H_2OH$	7.0	36.3	2.0×10^2	1.8×10^3
54	2,6-$[(CH_3)_3C]_2$-4-$C_6H_5CH_2C_6H_2OH$	1.2	33.4	5.8×10^2	4.3×10^3
55	2,6-$[(CH_3)_3C]_2$-4-$(CH_3)_3COC_6H_2OH$	−7.2	29.2	2.6×10^3	1.5×10^4
56	2,6-$[(CH_3)_3C]_2$-4-$(CH_3)_3COC(O)C_6H_2OH$	9.2	37.5	1.3×10^2	1.3×10^3
57	2,6-$[(CH_3)_3C]_2$-4-$HOOCC_6H_2OH$	10.3	38.0	1.1×10^2	1.1×10^3
58	2,6-$[(CH_3)_3C]_2$-4-ClC_6H_2OH	6.0	35.8	2.4×10^2	2.1×10^3
60	2,6-$[(CH_3)_3C]_2$-4-CNC_6H_2OH	13.9	39.9	55.1	6.2×10^2
61	2,6-$[(CH_3)_3C]_2$-4-$CHOC_6H_2OH$	9.3	37.5	1.3×10^2	1.3×10^3
62	2,6-$[(CH_3)_3C]_2$-4-$CH_3OC_6H_2OH$	−7.8	28.9	2.9×10^3	1.7×10^4
68	2,6-$[(CH_3)_3C]_2$-4-$NO_2C_6H_2OH$	19.5	43.0	18.0	2.4×10^2
70	2,6-$[(CH_3)_3C]_2$-4-$C_6H_5C_6H_2OH$	−0.8	32.4	8.2×10^2	5.9×10^3
73	2,6-$[(CH_3)_3C]_2$-4-$C_6H_5SC_6H_2OH$	7.9	36.8	1.7×10^2	1.6×10^3
89	2,6-$(C_6H_5)_2$-4-$NH_2CH_2C_6H_2OH$	−15.6	25.2	1.1×10^4	5.1×10^4
91	2,6-$(C_6H_5)_2$-4-$HOOCCH_2C_6H_2OH$	−12.6	26.6	6.7×10^3	3.4×10^4
94	2,6-$(C_6H_5)_2$-4-$CH_3SC_6H_2OH$	1.9	33.7	5.2×10^2	4.0×10^3
95	2,6-$(C_6H_5)_2$-4-$CH_3OC_6H_2OH$	−10.5	27.6	4.7×10^3	2.5×10^4

No.	Phenol	ΔH/ kJ mol^{-1}	E/ kJ mol^{-1}	k (333 K)/ l mol^{-1} s^{-1}	k (400 K)/ l mol^{-1} s^{-1}
122	$Si[2,6-[(CH_3)_3C]_2-4-OCH_2CH_2CH_2C_6H_2OH]_4$	2.4	34.0	4.6×10^2	3.6×10^3
123	$4,4'-CH_2SCH_2-[2,6-[(CH_3)_3C]_2C_6H_2OH]_2$	1.4	33.5	5.6×10^2	4.2×10^3
	cyclo-$[C(CH_3)_2CH_2C(O)CH_2C(CH_3)_2N(O^\bullet)]$				
1	$1,3,5-[3',5'-[(CH_3)_2C]_2-4'-HO-C_6H_5CH_2]_3C_6H_3$	44.5	57.4	9.9×10^{-2}	3.19
5	$2-HO-3,5-[(CH_3)_3C]_2C_6H_2OH$	37.9	53.4	0.42	10.6
24	$2,2'-CH_2-[4,6-[(CH_3)_3C]_2C_6H_2OH]_2$	40.5	54.9	0.24	6.77
31	$C-[4-(CH_2)_2COOCH_2-2,6-[(CH_3)_3C]_2C_6H_2OH]$	40.6	55.0	0.24	6.57
49	$2,4-[(CH_3)_3C]_2-6-CH_3-C_6H_2OH$	53.5	62.9	1.4×10^{-2}	0.61
50	$4,4'-[2,6-[(CH_3)_3C]_2C_6H_2OH]_2$	42.2	56.0	0.16	4.86
51	$2,6-[(CH_3)_3C]_2C_6H_3OH$	46.3	58.4	6.9×10^{-2}	2.36
52	$2,6-[(CH_3)_3C]_2-4-CH_3CONHC_6H_2OH$	26.3	46.7	4.70	79.7
53	$2,6-[(CH_3)_3C]_2-4-CH_3COC_6H_2OH$	45.1	57.7	8.9×10^{-2}	2.9
54	$2,6-[(CH_3)_3C]_2-4-C_6H_5CH_2C_6H_2OH$	39.3	54.2	0.31	8.35
55	$2,6-[(CH_3)_3C]_2-4-(CH_3)_3COC_6H_2OH$	30.9	49.3	1.85	36.5
56	$2,6-[(CH_3)_3C]_2-4-(CH_3)_3COC(O)C_6H_2OH$	47.3	59.0	5.6×10^{-2}	1.97
57	$2,6-[(CH_3)_3C]_2-4-HOOCC_6H_2OH$	48.4	59.7	4.3×10^{-2}	1.60
58	$2,6-[(CH_3)_3C]_2-4-ClC_6H_2OH$	44.1	57.1	0.11	3.49
60	$2,6-[(CH_3)_3C]_2-4-CNC_6H_2OH$	52.0	62.0	1.9×10^{-2}	0.80
61	$2,6-[(CH_3)_3C]_2-4-CHOC_6H_2OH$	47.4	59.1	5.4×10^{-2}	1.91
62	$2,6-[(CH_3)_3C]_2-4-CH_3OC_6H_2OH$	30.3	49.0	2.06	39.9
68	$2,6-[(CH_3)_3C]_2-4-NO_2C_6H_2OH$	57.6	65.5	5.3×10^{-3}	0.28
70	$2,6-[(CH_3)_3C]_2-4-C_6H_5C_6H_2OH$	37.3	53.0	0.49	12.0
73	$2,6-[(CH_3)_3C]_3-4-C_6H_5SC_6H_2OH$	46.0	58.3	7.2×10^{-2}	2.43
89	$2,6-(C_6H_5)-4-NH_2C_6H_2OH$	22.5	44.6	10.0	1.5×10^{-2}
91	$2,6-(C_6H_5)-4-HOOCCH_2C_6H_2OH$	25.5	46.3	5.46	90.0
94	$2,6-(C_6H_5)_2-4-CH_3SC_6H_2OH$	40.0	54.6	0.27	7.41
95	$2,6-(C_6H_5)_2-4-CH_3OC_6H_2OH$	27.6	47.5	3.54	62.6
122	$Si[2,6-[(CH_3)_3C]_2-4-OCH_2CH_2C_6H_2OH$	40.5	54.9	0.24	67.7

No.	Phenol	ΔH/ kJ mol^{-1}	E/ kJ mol^{-1}	k (333 K)/ l mol^{-1} s^{-1}	k (400 K)/ l mol^{-1} s^{-1}
123	4,4'-CH_2SCH_2-[2,6-[$(CH_3)_3C$]$_2C_6H_2OH$]$_2$	39.5	54.3	0.30	8.11
	cyclo-[$C(CH_3)_2C(CH_3)_2N{=}C(C_6H_5)N(O^•)$]				
1	1,3,5-[3',5'-[$(CH_3)_2C$]$_2$-4'-HO-$C_6H_5CH_2$]$_3C_6H_3$	32.7	50.4	1.24	26.2
5	2-HO-3,5-[$(CH_3)_3C$]$_2C_6H_2OH$	26.1	46.6	4.90	82.1
24	2,2'-CH_2-[4,6-[$(CH_3)_3C$]$_2C_6H_2OH$]$_2$	28.7	48.1	2.85	52.3
31	C-[4-$(CH_2)_2COOCH_2$-2,6-[$(CH_3)_3C$]$_2C_6H_2OH$]	28.8	48.1	2.85	52.3
49	2,4-[$(CH_3)_3C$]$_2$-6-CH_3-C_6H_2OH	41.7	55.7	0.18	5.32
50	4,4'-[2,6-[$(CH_3)_3C$]$_2C_6H_2OH$]$_2$	30.4	49.0	2.06	40.0
51	2,6-[$(CH_3)_3C$]$_2C_6H_3OH$	34.5	51.4	0.86	19.4
52	2,6-[$(CH_3)_3C$]$_2$-4-$CH_3CONHC_6H_2OH$	14.5	40.3	47.7	5.5×10^2
53	2,6-[$(CH_3)_3C$]$_2$-4-$CH_3COC_6H_2OH$	33.3	50.7	1.11	23.9
54	2,6-[$(CH_3)_3C$]$_2$-4-$C_6H_5CH_2C_6H_2OH$	27.5	47.4	3.67	64.6
55	2,6-[$(CH_3)_3C$]$_2$-4-$(CH_3)_3COC_6H_2OH$	19.1	42.7	20.0	2.6×10^2
56	2,6-[$(CH_3)_3C$]$_2$-4-$(CH_3)_3COC(O)C_6H_2OH$	35.5	52.0	0.70	16.2
57	2,6-[$(CH_3)_3C$]$_2$-4-$HOOCC_6H_2OH$	36.6	52.6	0.56	13.5
58	2,6-[$(CH_3)_3C$]$_2$-4-ClC_6H_2OH	32.3	50.1	1.38	28.7
60	2,6-[$(CH_3)_3C$]$_2$-4-CNC_6H_2OH	40.2	54.8	0.25	7.0
61	2,6-[$(CH_3)_3C$]$_2$-4-$CHOC_6H_2OH$	35.6	52.1	0.67	15.7
62	2,6-[$(CH_3)_3C$]$_2$-4-$CH_3OC_6H_2OH$	18.5	42.4	0.22	2.9×10^2
68	2,6-[$(CH_3)_3C$]$_2$-4-$NO_2C_6H_2OH$	45.8	58.1	7.7×10^{-2}	2.59
70	2,6-[$(CH_3)_3C$]$_2$-4-$C_6H_5C_6H_2OH$	25.5	46.3	5.46	89.9
73	2,6-[$(CH_3)_3C$]$_2$-4-$C_6H_5SC_6H_2OH$	34.2	51.2	0.93	20.6
89	2,6-$(C_6H_5)_2$-4-$NH_2CH_2C_6H_2OH$	10.7	38.2	1.0×10^2	1.0×10^3
91	2,6-$(C_6H_5)_2$-4-$HOOCCH_2C_6H_2OH$	13.7	39.8	57.1	6.3×10^2
94	2,6-$(C_6H_5)_2$-4-$CH_3SC_6H_2OH$	28.2	47.8	4.24	57.2
95	2,6-$(C_6H_5)_2$-4-$CH_3OC_6H_2OH$	15.8	40.9	38.4	4.6×10^2
122	Si[2,6-[$(CH_3)_3C$]$_2$-4-$OCH_3CH_2C_6H_2OH$]$_4$	28.7	48.1	2.85	52.3
123	4,4'-CH_2SCH_2-[2,6-[$(CH_3)_3C$]$_2C_6H_2OH$]$_2$	27.7	47.6	3.41	60.8

No.	Phenol	ΔH/ kJ mol^{-1}	E/ kJ mol^{-1}	k (333 K)/ l mol^{-1} s^{-1}	k (400 K)/ l mol^{-1} s^{-1}
	cyclo-[(C(CH$_3$)$_2$CH(CH$_3$)N(CH$_3$)$_2$N(O$^•$)]				
1	1,3,5-[3′,5′-[(CH$_3$)$_3$C]$_2$-4′-HO-C$_6$H$_5$CH$_2$]$_3$C$_6$H$_3$	54.1	63.3	1.2×10^{-2}	0.54
5	2-HO-3,5-[(CH$_3$)$_3$C]$_2$C$_6$H$_2$OH	47.5	59.2	5.2×10^{-2}	1.86
24	2,2′-CH$_2$-[4,6-[(CH$_3$)$_3$C]$_2$C$_6$H$_2$OH]$_2$	50.1	60.8	2.9×10^{-2}	1.15
31	C-[4-(CH$_2$)$_2$COOCH$_2$-2,6-[(CH$_3$)$_3$C]$_2$C$_6$H$_2$OH]	50.2	60.8	2.9×10^{-2}	1.15
49	2,4-[(CH$_3$)$_3$C]$_2$-6-CH$_3$-C$_6$H$_2$OH	63.1	69.0	1.5×10^{-3}	9.8×10^{-2}
50	4,4′-[2,6-[(CH$_3$)$_3$C]$_2$C$_6$H$_2$OH]$_2$	51.8	61.8	2.0×10^{-2}	0.85
51	2,6-[(CH$_3$)$_3$C]$_2$C$_6$H$_3$OH	55.9	64.4	7.9×10^{-3}	0.39
52	2,6-[(CH$_3$)$_3$C]$_2$-4-CH$_3$CONHC$_6$H$_2$OH	35.9	52.2	0.65	15.2
53	2,6-[(CH$_3$)$_3$C]$_2$-4-CH$_3$COC$_6$H$_2$OH	54.7	63.6	1.1×10^{-2}	0.49
54	2,6-[(CH$_3$)$_3$C]$_2$-4-C$_6$H$_5$CH$_2$C$_6$H$_2$OH	48.9	60.0	3.9×10^{-2}	1.46
55	2,6-[(CH$_3$)$_3$C]$_2$-4-(CH$_3$)$_3$COC$_6$H$_2$OH	40.5	54.9	0.24	6.77
56	2,6-[(CH$_3$)$_3$C]$_2$-4-(CH$_3$)$_3$COC(O)C$_6$H$_2$OH	56.9	65.0	6.4×10^{-3}	0.32
57	2,6-[(CH$_3$)$_3$C]$_2$-4-HOOCC$_6$H$_2$OH	58.0	65.7	4.9×10^{-3}	0.26
58	2,6-[(CH$_3$)$_3$C]$_2$-4-ClC$_6$H$_2$OH	53.7	63.0	1.3×10^{-2}	0.59
60	2,6-[(CH$_3$)$_3$C]$_2$-4-CNC$_6$H$_2$OH	61.6	68.0	2.2×10^{-3}	0.13
61	2,6-[(CH$_3$)$_3$C]$_2$-4-CHOC$_6$H$_2$OH	57.0	65.1	6.1×10^{-3}	0.32
62	2,6-[(CH$_3$)$_3$C]$_2$-4-CH$_3$OC$_6$H$_2$OH	39.9	54.6	0.27	7.41
68	2,6-[(CH$_3$)$_3$C]$_2$-4-NO$_2$C$_6$H$_2$OH	67.2	71.7	5.7×10^{-4}	4.3×10^{-2}
70	2,6-[(CH$_3$)$_3$C]$_2$-4-C$_6$H$_5$C$_6$H$_2$OH	46.9	58.8	6.0×10^{-2}	2.10
73	2,6-[(CH$_3$)$_3$C]$_2$-4-C$_6$H$_5$SC$_6$H$_2$OH	55.6	64.2	8.5×10^{-3}	0.41
89	2,6-(C$_6$H$_5$)$_2$-4-NH$_2$CH$_2$C$_6$H$_2$OH	32.1	50.0	1.43	29.5
91	2,6-(C$_6$H$_5$)$_2$-4-HOOCCH$_2$C$_6$H$_2$OH	35.1	51.8	0.75	17.2
94	2,6-(C$_6$H$_5$)$_2$-4-CH$_3$SC$_6$H$_2$OH	49.6	60.5	3.2×10^{-2}	1.26
95	2,6-(C$_6$H$_5$)$_2$-4-CH$_3$OC$_6$H$_2$OH	37.2	53.0	0.49	12.0
122	Si[2,6-[(CH$_3$)$_3$C]$_2$-4-OCH$_2$CH$_2$C$_6$H$_2$OH]$_4$	50.1	60.8	2.9×10^{-2}	1.15
123	4,4′-CH$_2$SCH$_2$-[2,6-[(CH$_3$)$_3$C]$_2$C$_6$H$_2$OH]$_2$	49.1	60.2	3.6×10^{-2}	1.38

Table 5.13
Enthalpies, activation energies and rate constants of reaction of nitroxyl radicals with aromatic amines calculated by formulas 1.15–1.17 and 1.20. The values of A, br_e and α, see Table 1.6

No.	Amine	ΔH/ kJ mol^{-1}	E/ kJ mol^{-1}	k (300 K)/ l mol^{-1} s^{-1}	k (400 K)/ l mol^{-1} s^{-1}
	$[(CH_3)_3C]_2C{=}NO^{\bullet}$				
135	$(2\text{-}C_{10}H_7)_2NH$	15.0	21.4	4.4×10^4	1.6×10^5
136	$(C_6H_5)_2NH$	26.2	27.5	4.9×10^3	2.6×10^4
137	$(4\text{-}BrC_6H_4)_2NH$	10.0	18.8	1.1×10^5	3.5×10^5
138	$(4\text{-}CH_3OC_6H_4)_2NH$	2.7	15.2	4.1×10^5	1.0×10^6
139	$(4\text{-}CH_3C_6H_4)_2NH$	12.1	19.9	7.6×10^4	2.5×10^5
140	$(4\text{-}(CH_3)_3CC_6H_4)_2NH$	6.9	17.2	2.0×10^5	5.7×10^5
141	$4\text{-}CH_3OC_6H_4NHC_6H_5$	8.7	18.1	1.4×10^5	4.3×10^5
142	$4\text{-}NO_2C_6H_4NHC_6H_4$	30.0	30.0	2.0×10^3	1.2×10^4
143	$4\text{-}(CH_3)_3CC_6H_4NHC_6H_5$	22.3	25.3	1.1×10^4	5.0×10^4
144	$1\text{-}C_{10}H_7NH_2$	36.2	36.2	4.2×10^2	3.8×10^3
145	$2\text{-}C_{10}H_7NH_2$	41.0	41.0	74.0	8.8×10^2
146	$2\text{-}C_{10}H_7NHC_6H_4NH\text{-}2\text{-}C_{10}H_7$	8.1	17.8	3.2×10^5	9.4×10^5
147	$4\text{-}C_8H_{17}NHC_6H_4NHC_8H_{17}$	8.4	18.0	3.0×10^5	9.0×10^5
148	$4\text{-}C_6H_5NHC_6H_4NHC_6H_5$	8.4	18.0	3.0×10^5	9.0×10^5
149	$4\text{-}(CH_3)_2CHC_6H_4NH\text{-}4\text{-}C_6H_4NHC_6H_4CH(CH_3)_2$	− 4.9	11.6	3.0×10^6	6.2×10^6
150	$4\text{-}C_6H_5NHC_6H_4NHCH(CH_3)_3$	1.7	14.7	4.9×10^5	1.2×10^6
151	$C_6H_5NH\text{-}1\text{-}3,7\text{-}[(CH_3)C]_2C_{10}H_5$	6.4	17.0	2.2×10^5	6.0×10^5
152	$1\text{-}C_{10}H_7NHC_6H_5$	13.4	20.6	5.9×10^4	2.0×10^5
153	$2\text{-}C_{10}H_7NHC_6H_5$	15.4	21.6	4.1×10^4	1.5×10^5
154	$2\text{-}C_{10}H_7NH(4'\text{-}C_6H_5O)C_6H_4$	11.0	19.3	9.4×10^4	3.0×10^5
155	1,2-*cyclo*-[C(*cyclo*-CH_2CH_2O)CH_2C(NH-(*cyclo*-C_6H_{11}))]C_6H_4	22.3	25.3	1.1×10^4	5.0×10^4
156	$1\text{-}[NHC_6H_4\text{-}4'\text{-}C(CH_3)_3]\text{-}4\text{-}(CH_3)_3CC_{10}H_6$	13.6	20.7	5.7×10^4	2.0×10^5
157	1,2-*cyclo*-[$C(CH_3)CH_2C(CH_3)_2NH$]C_6H_4	21.0	24.6	1.4×10^4	6.1×10^4

No.	Amine	ΔH/ kJ mol^{-1}	E/ kJ mol^{-1}	k (300 K)/ l mol^{-1} s^{-1}	k (400 K)/ l mol^{-1} s^{-1}
	cyclo-[C(CH$_3$)$_2$CH$_2$C(O)CH$_2$C(CH$_3$)$_2$N(O$^{\bullet}$)]				
135	(2-C$_{10}$H$_7$)$_2$NH	53.1	53.1	0.47	11.6
136	(C$_6$H$_5$)$_2$NH	64.3	64.3	8.2×10^{-3}	0.40
137	(4-BrC$_6$H$_4$)$_2$NH	48.1	48.1	2.85	52.3
138	(4-CH$_3$OC$_6$H$_4$)$_2$NH	40.8	40.8	39.8	4.7×10^2
139	(4-CH$_3$C$_6$H$_4$)$_2$NH	50.2	50.2	1.33	27.8
140	(4-(CH$_3$)$_3$CC$_6$H$_4$)$_2$NH	45.0	45.0	8.73	1.3×10^2
141	4-CH$_3$OC$_6$H$_4$NHC$_6$H$_5$	46.8	46.8	4.56	0.77
142	4-NO$_2$C$_6$H$_4$NHC$_6$H$_4$	68.1	68.1	2.1×10^{-3}	0.13
143	4-(CH$_3$)$_3$CC$_6$H$_4$NHC$_6$H$_5$	60.4	60.4	3.4×10^{-2}	1.29
144	1-C$_{10}$H$_7$NH$_2$	74.3	74.3	4.4×10^{-4}	4.0×10^{-2}
145	2-C$_{10}$H$_7$NH$_2$	79.1	79.1	7.8×10^{-5}	9.4×10^{-3}
146	2-C$_{10}$H$_7$NHC$_6$H$_4$NH-2-C$_{10}$H$_7$	46.2	46.2	11.3	1.9×10^2
147	4-C$_8$H$_{17}$NHC$_6$H$_4$NHC$_8$H$_{17}$	46.5	46.5	10.2	1.7×10^2
148	4-C$_6$H$_5$NHC$_6$H$_4$NHC$_6$H$_5$	46.5	46.5	10.2	1.7×10^2
149	4-(CH$_3$)$_2$CHC$_6$H$_4$NH-4-C$_6$H$_4$NHC$_6$H$_4$CH(CH$_3$)$_2$	33.2	32.2	1.8×10^3	1.2×10^4
150	4-C$_6$H$_5$NHC$_6$H$_4$NHCH(CH$_3$)$_2$	39.8	39.8	57.1	6.3×10^2
151	C$_6$H$_5$NH-1-3,7-[(CH$_3$)$_3$C]$_2$C$_{10}$H$_5$	44.5	44.5	10.5	1.5×10^2
152	1-C$_{10}$H$_7$NHC$_6$H$_5$	51.5	51.5	0.83	18.8
153	2-C$_{10}$H$_7$NHC$_6$H$_5$	53.5	53.5	0.41	10.3
154	2-C$_{10}$H$_7$NH(4'-C$_6$H$_5$O)C$_6$H$_4$	49.1	49.1	1.98	38.7
155	1,2-*cyclo*-[C(*cyclo*-CH$_2$CH$_2$O)CH$_2$C(NH-*cyclo*-C$_6$H$_{11}$))]C$_6$H$_4$	60.4	60.4	3.4×10^{-2}	1.30
156	1-[NHC$_6$H$_4$-4'-C(CH$_3$)$_3$]-4-(CH$_3$)$_3$CC$_{10}$H$_6$	51.7	51.7	0.78	17.7
157	1,2-*cyclo*-[C(CH$_3$)CH$_2$C(CH$_3$)$_2$NH]C$_6$H$_4$	59.1	59.1	5.4×10^{-2}	1.91

No.	Amine	ΔH/ kJ mol^{-1}	E/ kJ mol^{-1}	k (300 K)/ l mol^{-1} s^{-1}	k (400 K)/ l mol^{-1} s^{-1}
	cyclo-[$C(CH_3)_2C(CH_3)_2N{=}C(C_6H_5)N(O^•)$]				
135	$(2\text{-}C_{10}H_7)_2NH$	41.3	41.3	33.2	4.0×10^2
136	$(C_6H_5)_2NH$	52.2	52.2	0.65	15.2
137	$(4\text{-}BrC_6H_4)_2NH$	36.3	36.3	2.0×10^2	1.8×10^3
138	$(4\text{-}CH_3OC_6H_4)_2NH$	29.0	29.1	2.7×10^3	1.6×10^4
139	$(4\text{-}CH_3C_6H_4)_2NH$	38.4	38.4	96	9.7×10^2
140	$(4\text{-}(CH_3)_3CC_6H_4)_2NH$	33.2	33.2	6.2×10^2	4.6×10^2
141	$4\text{-}CH_3OC_6H_4NHC_6H_5$	35.0	35.0	3.2×10^2	2.7×10^7
142	$4\text{-}NO_2C_6H_4NHC_6H_4$	56.3	56.3	0.15	4.44
143	$4\text{-}(CH_3)_3CC_6H_4NHC_6H_5$	48.6	48.6	2.38	45.0
144	$1\text{-}C_{10}H_7NH_2$	62.5	62.5	3.1×10^{-2}	1.38
145	$2\text{-}C_{10}H_7NH_2$	67.3	67.3	5.5×10^{-3}	0.33
146	$2\text{-}C_{10}H_7NHC_6H_4NH\text{-}2\text{-}C_{10}H_7$	34.4	34.4	8.0×10^2	6.4×10^3
147	$4\text{-}C_8H_{17}NHC_6H_4NHC_8H_{17}$	34.7	34.7	7.2×10^2	2.9×10^3
148	$4\text{-}C_6H_5NHC_6H_4NHC_6H_5$	34.7	34.7	7.2×10^2	2.9×10^3
149	$4\text{-}(CH_3)_2CHC_6H_4NH\text{-}4\text{-}C_6H_4NHC_6H_4CH(CH_3)_2$	21.4	24.8	2.6×10^4	1.2×10^5
150	$4\text{-}C_6H_5NHC_6H_4NHCH(CH_3)_2$	28.0	28.5	3.4×10^3	1.9×10^4
151	$C_6H_5NH\text{-}1\text{-}3,7\text{-}[(CH_3)_3C]_2C_{10}H_5$	32.7	32.7	7.4×10^2	5.4×10^3
152	$1\text{-}C_{10}H_7NHC_6H_5$	39.7	39.7	59.2	6.5×10^2
153	$2\text{-}C_{10}H_7NHC_6H_5$	41.7	41.7	28.8	3.6×10^2
154	$2\text{-}C_{10}H_7NH(4'\text{-}C_6H_5O)C_6H_4$	37.3	37.3	1.4×10^2	1.3×10^3
155	1,2-*cyclo*-[C(*cyclo*-$CH_2CH_2O)CH_2C(NH$-*cyclo*-$C_6H_{11}))]C_6H_4$	48.6	48.6	2.38	45.0
156	$1\text{-}[NHC_6H_4\text{-}4'\text{-}C[(CH_3)_3]\text{-}4\text{-}(CH_3)_3CC_{10}H_6$	39.9	39.9	55.1	6.2×10^2
157	1,2-*cyclo*-[$C(CH_3)CH_2C(CH_3)_2NH]C_6H_4$	47.3	47.3	3.80	66.5

Table 5.14
Rate constants of addition reaction of peroxyl radicals to double bond of nitrones

No.	Nitrone	$RO_2^•$	T/ K	k/ l mol^{-1} s^{-1} or log k = $A - E/\theta$	Ref.
238	N-(benzylidene)-aniline-N-oxide; $C_6H_5CH{=}N(O)C_6H_5$	$CH_3CH(O_2^•)(CH_2)_{11}CH_3$	323–353	8.7 – 31.2/θ	23
239	N-(2-Phehylamino-2-oxoethylidene)aniline-N-oxide; $C_6H_5NHC(O)CH{=}N(O)C_6H_5$	$CH_3CH(O_2^•)(CH_2)_{11}CH_3$	323–353	9.2 – 37.2/θ	23
239	N-(2-Phehylamino-2-oxoethylidene)aniline-N-oxide; $C_6H_5NHC(O)CH{=}N(O)C_6H_5$	$CH_3C(O)OCH(O_2^•)CH_2OC(O)CH_3$	323–343	12.4 – 48.0/θ	24
239	N-(2-Phehylamino-2-oxoethylidene)aniline-N-oxide; $C_6H_5NHC(O)CH{=}N(O)C_6H_5$	$C_6H_5CH(O_2^•)OC(O)CH_3$	383–408	10.5 – 39.5/θ	24
240	N-(2-Phenylamino-2-oxoethylidene)-*p*-bromaniline-N-oxide; $C_6H_5NHC(O)CH{=}N(O)\text{-}4\text{-}BrC_6H_4$	$CH_3CH(O_2^•)(CH_2)_{11}CH_3$	333	6.6×10^3	25
241	N-(2-Phenylamino-2-oxoethylidene)-*p*-dibutylaminoaniline-N-oxide; $C_6H_5NHC(O)CH{=}N(O)\text{-}4\text{-}(C_4H_9)_2NC_6H_4$	$CH_3CH(O_2^•)(CH_2)_{11}CH_3$	333	1.0×10^3	25
241	N-(2-Phenylamino-2-oxoethylidene)-*p*-dibutylaminoaniline-N-oxide; $C_6H_5NHC(O)CH{=}N(O)\text{-}4\text{-}(C_4H_9)_2NC_6H_4$	$CH_3C(O)OCH(O_2^•)CH_2OC(O)CH_3$	327 – 358	9.6 – 30.9/θ	24
242	N-(2-Phenylamino-2-oxoethylidene)-*p*-diethylaminoanilie-N-oxide; $C_6H_4NHC(O)CH{=}N(O)\text{-}4\text{-}(C_2H_5)_2NC_6H_4$	$CH_3C(O)OCH(O_2^•)CH_2OC(O)CH_3$	327–343	13.1 – 56.0/θ	24
243	N-(2-Phenylamino-2-oxoethylidene)-*o*-methylaniline-N-oxide; $C_6H_4NHC(O)CH{=}N(O)\text{-}2\text{-}CH_3C_6H_4$	$CH_3CH(O_2^•)(CH_2)_{11}CH_3$	333	1.7×10^3	25
243	N-(2-Phenylamino-2-oxoethylidene)-*o*-methylaniline-N-oxide; $C_6H_4NHC(O)CH{=}N(O)\text{-}2\text{-}CH_3C_6H_4$	$CH_3C(O)OCH(O_2^•)CH_2OC(O)CH_3$	338–360	12.2 – 50.5/θ	24

No.	Nitrone	$RO_2^{\bullet}$	T/ K	k/ l mol^{-1} s^{-1} or log k = $A - E/\theta$	Ref.
244	N-(2-Phenylamino-2-oxoethylidene)-*p*-methylaniline-N-oxide; $C_6H_4NHC(O)CH{=}N(O)\text{-}4\text{-}CH_3C_6H_4$	$CH_3CH(O_2^{\bullet})(CH_2)_{11}CH_3$	333	3.8×10^3	25
244	N-(2-Phenylamino-2-oxoethylidene)-*p*-methylaniline-N-oxide; $C_6H_4NHC(O)CH{=}N(O)\text{-}4\text{-}CH_3C_6H_4$	$CH_3C(O)CH(O_2^{\bullet})CH_2OC(O)CH_3$	333–349	$10.0 - 32.1/\theta$	24
245	N-(2-Phenylamino-2-oxoethylidene)-*p*-dimethylaminoaniline-N-oxide; $C_6H_4NHC(O)CH{=}N(O)\text{-}4\text{-}(CH_3)_2NC_6H_4$	$CH_3CH(O_2^{\bullet})(CH_2)_{11}CH_3$	333	9.7×10^2	25
246	N-(2-Phenylamino-2-oxoethylidene)-*p*-propyloxyaniline-N-oxide; $C_6H_4NHC(O)CH{=}N(O)\text{-}4\text{-}C_3H_7OC_6H_4$	$CH_3CH(O_2^{\bullet})(CH_2)_{11}CH_3$	333	7.7×10^2	25
246	N-(2-Phenylamino-2-oxoethylidene)-*p*-propyloxyaniline-N-oxide; $C_6H_4NHC(O)CH{=}N(O)\text{-}4\text{-}C_3H_7OC_6H_4$	$CH_3C(O)OCH(O_2^{\bullet})CH_2OC(O)CH_3$	328–335	$12.4 - 48.2/\theta$	24

Table 5.15
Rate constants of reaction of hydroxylamines with oxygen: $AmOH + O_2 \rightarrow AmO^{\bullet} + HO_2^{\bullet}$, calculated by formula 1.17 with $A = 3 \times 10^{10}$ l mol^{-1} s^{-1} per one O—H-bond and $E = \Delta H$

No.	Hydroxylamine	ΔE/ kJ mol^{-1}	k (400 K)/ l mol^{-1} s^{-1}	k (500 K)/ l mol^{-1} s^{-1}
158	*cyclo*-[-CH(NOH)(CH$_2$)$_3$-CH(CH$_2$)$_3$]	115.4	8.5×10^{-6}	8.8×10^{-3}
159	*cyclo*-[-CH(NOH)(CH$_2$)$_2$-CH(CH$_2$)$_3$]	118.6	3.3×10^{-6}	4.1×10^{-3}
160	*cyclo*-[C(CH$_3$)$_2$CH$_2$C(O)NHCH$_2$C(CH$_3$)$_2$N(OH)]	100.9	6.7×10^{-4}	0.29
163	*cyclo*-[C(CH$_3$)$_2$C(4′-FC$_6$H$_4$)=NC(CH$_3$)$_2$N(OH)]	92.7	7.9×10^{-3}	2.1
166	*cyclo*-[C(CH$_3$)$_2$C(CH$_3$)=NC(CH$_3$)$_2$N(OH)]	91.8	1.0×10^{-2}	2.6
167	*cyclo*-[C(CH$_3$)$_2$C(4′-CH$_3$C$_6$H$_4$)=NC(CH$_3$)$_2$N(OH)]	93.5	6.2×10^{-3}	1.7
172	*cyclo*-[C(CH$_3$)$_2$C(CH$_3$)$_2$N=C(C$_6$H$_5$)N(OH)]	108.8	6.2×10^{-5}	4.3×10^{-2}
173	*cyclo*-[C(CH$_3$)$_2$C(C$_6$H$_5$)=N(O)C(CH$_3$)[N(CH$_3$)OH]N(OH)]	106.6	1.2×10^{-4}	7.3×10^{-2}
174	*cyclo*-[C(CH$_3$)$_2$C(CH$_3$)=N(O)C(CH$_3$)$_2$N(OH)]	102.4	4.2×10^{-4}	0.20

No.	Hydroxylamine	ΔE/ kJ mol^{-1}	k (400 K)/ l mol^{-1} s^{-1}	k (500 K)/ l mol^{-1} s^{-1}
178	*cyclo*-[$C(CH_3)_2C(CHBr_2){=}N(O)C(CH_3)_2N(OH)$]	108.0	7.9×10^{-5}	5.2×10^{-2}
183	*cyclo*-[$C(CH_3)_2C(CH_2Br){=}N(O)C(CH_3)_2N(OH)$]	102.9	3.7×10^{-4}	0.18
185	*cyclo*-[$C(CH_3)_2C(4'\text{-}CH_3OC_6H_4){=}N(O)C(CH_3)_2N(OH)$]	99.8	9.2×10^{-4}	0.38
200	1,2-$C_6H_3[C(C_6H_5)_3]$-*cyclo*-[$C(CH_3){=}CHC(CH_3)_2N(OH)$]	86.1	5.7×10^{-2}	10.1
201	$C_8H_{13}[(CH_3)_3C]NOH$	131.3	7.2×10^{-8}	1.9×10^{-4}
202	$C_6H_5C(O)[(CH_3)_3C]NOH$	119.6	2.4×10^{-6}	3.2×10^{-3}
207	$[(CH_3)_3C]_2C{=}NOH$	135.1	2.3×10^{-8}	7.7×10^{-5}
208	$(CF_3)_2NOH$	142.3	2.6×10^{-9}	1.4×10^{-5}
209	$(CH_3)_2CHN(OH)C(CH_3)_3$	149.3	3.2×10^{-10}	2.5×10^{-6}
211	$(CH_3)_2CHCH_2C(O)N(OH)C(CH_3)_3$	113.1	1.7×10^{-5}	1.5×10^{-2}
212	$4\text{-}NO_2C_6H_4N(OH)C(CH_3)_3$	132.2	5.5×10^{-8}	1.5×10^{-4}
213	$C_6H_5CH{=}CHC(O)N(OH)C(CH_3)_3$	118.0	3.9×10^{-6}	4.7×10^{-3}
214	$4\text{-}C_6H_5C_6H_4C(O)N(OH)C(CH_3)_3$	121.0	1.6×10^{-6}	2.3×10^{-3}
215	$C_6H_5CH_2CH_2C(O)N(OH)C(CH_3)_3$	115.2	9.1×10^{-6}	9.2×10^{-3}
216	$CH_3(CH_2)_9C(O)N(OH)C(CH_3)_3$	113.8	1.4×10^{-5}	1.3×10^{-2}
217	*cyclo*-[$C(CH_3)_2(CH_2)_3C(CH_3)_2N(OH)$]	87.8	3.4×10^{-2}	6.7
218	*cyclo*-[$C(CH_3)_2CH_2CH(OC(O)C_6H_5)CH_2C(CH_3)_2N(OH)$]	96.9	2.2×10^{-3}	0.75
219	*cyclo*-[$C(CH_3)_2CHBrC(O)CH_2C(CH_3)_2N(OH)$]	100.0	8.7×10^{-4}	0.36
220	*cyclo*-[$C(CH_3)_2CHClC(O)CH_2C(CH_3)_2N(OH)$]	104.5	2.3×10^{-4}	0.12
221	*cyclo*-[$C(CH_3)_2CH_2CH(OH)CH_2C(CH_3)_2N(OH)$]	98.8	1.3×10^{-3}	0.48
222	*cyclo*-[$C(CH_3)_2CH_2C(O)CH_2C(CH_3)_2N(OH)$]	97.0	2.2×10^{-3}	0.74
223	*cyclo*-[$C(CH_3)_2CH(C(O)NH_2)CH_2C(CH_3)_2N(OH)$]	87.6	3.6×10^{-2}	7.1
224	*cyclo*-[$C(CH_3)_2CH(C(O)OH)CH_2C(CH_3)_2N(OH)$]	86.7	4.8×10^{-2}	8.8
225	*cyclo*-[$C(CH_3)_2CH(OH)CH_2C(CH_3)_2N(OH)$]	85.7	6.4×10^{-2}	11.1
226	*cyclo*-[$C(CH_3)_2C(O)CH_2C(CH_3)_2N(OH)$]	95.3	3.6×10^{-3}	1.1
227	*cyclo*-[$C(CH_3)_2C(O)N(OH)C(CH_3)_2N(OH)$]	97.5	1.9×10^{-3}	0.65
232	*cyclo*-[$C(CH_3)_2N(O){=}C(CH_3)CH_2C(CH_3)_2N(OH)$]	105.4	1.7×10^{-4}	9.8×10^{-2}
235	*cyclo*-[$C(CH_3)_2CH_2C(C_6H_5){=}CHC(CH_3)_2N(OH)$]	92.5	8.3×10^{-3}	2.2
237	1,2-$C_6H_3[C(C_6H_5)_3]$-*cyclo*-[$C(CH_3)(C_6H_5)CH_2C(CH_3)_2N(OH)$]	89.4	2.1×10^{-2}	4.6

REFERENCES

1 **Mahoney, L. R., Mendenhall G. D., Ingold K. U.**, Calorimetric and equilibrium studies on some stable nitroxide and iminoxy radicals. Approximate O—H-bond dissociation energies in hydroxylamines and oximes, *J. Am. Chem. Soc.*, 95, 8610, 1973.

2 **Dikanov, S. A., Grigoriev, I. A., Volodarsky, L. B., Tsvetkov, Yu. D.**, Oxidative activity of nitroxyl radicals in their reactions with sterically hindered hydroxylamines, *Zh. Fiz. Khim.*, 56, 2762, 1982.

3 **Denisov, E. T.**, The analysis of reactivity of nitroxyl radicals on the basis of parabolic model of transition state, *Kinet. Katal.*, 36, N 3, 1995.

4 **Malievky, A. D., Koroteev, S.V.**,Reactioon of hydrogen exchange between sterically-hyndered hydroxylamine and nitroxyl radical, *Izv. Russ. Akad. Nauk. Ser. Khim.* (in press).

5 **Jenkins, T. C., Perkins, M. J., Siew, N. P. Y.**, Structure and stability of acylnitroxyls, *J. Chem. Soc. Chem. Comm.*, 1975, 880.

6 **Doba, T., Ingold, K. U.**, Kinetic applications of electron paramagnetic resonance spectroscopy. 42. Some reactions of the bis(trifluoromethyl) aminoxyl radical, *J. Am. Chem. Soc.*, 106, 3958, 1984.

7 **Fischer H., Ed.** *LANDOLT-BORNSTEIN Numerical Data and Functional Relationships in Science and Technology, New Series*; Springer-Verlag, Berlin, 1984, V. 13, S/vol C,. p. 222–226.

8 **Nigam, S., Asmus, K. -D., Willson R. L.**, Electron transfer and addition reactions of free nitroxyl radicals with radiation induced radicals, *J. Chem. Soc. Faraday Trans. I*, 1976, 2314.

9 **Chateauneuf, J., Lusztyk, J., Ingold, K. U.**, Absolute rate constants for the reactions of some carbon-centered radicals with 2,2,6,6-tetramethylpiperidine-N-oxyl, *J. Org. Chem.*, 53, 1629, 1988.

10 **Beckwith, A. L. J., Bowry, V. W.**, Kinetics of coupling reactions of the nitroxyl radical 1,1,3,3,-tetramethylisoindoline-2-oxyl with carbon-centered radicals, *J. Org. Chem.*, 53, 1632, 1988.

11 **Koroli, L. L., Kuzmin, V. A., Khudyakov I. V.**, Kinetics of recombination, dismutation and disproportionation reactions involving neutral ketyl radicals and radical anions, *Int. J. Chem. Kinet.*, 16, 379, 1984.

12 **Bowman, D. F., Gillman, T., Ingold, K. U.**, Kinetic application of EPR spectroscopy, 3. Self-reactions of dialkyl nitroxide radicals, *J. Am. Chem. Soc.*, 91, 6555, 1971.

13 **Bowman, D. F., Brokenshire, J. L., Gillan, T., Ingold, K. U.**, Kinetic application of EPR spectroscopy. 2. Self-reactions of N-alkyl nitroxides and N-phenyl nitroxide, *J. Am. Chem. Soc.*, 93, 6551, 1971.

14 **Adamic, K., Bowman, D. F., Gillan, T., Ingold, K. U.**, Kinetic application of EPR spectroscopy. 1. Self-reaction of diethyl nitroxide radicals, *J. Am. Chem. Soc.*, 93, 902, 1971.

15 **Marriott, P.R., Ingold, K.U.**, A reexamination of the decay kinetics of diethyl nitroxide in aqueous solution by EPR spectroscopy, *J. Phys. Chem.*, 84, 937, 1980.

16 **Griller, D., Perkins, M. J.**, Unusual kinetic behavior in the self-reactions of acyl methyl nitroxides, *J. Am. Chem. Soc.*, 102, 1354, 1980.

17 **Brokenshire, J. L., Roberts, J. R., Ingold, K. U.**, Kinetic application of EPR spectroscopy. 7. Self reactions of iminoxy radicals, *J. Am. Chem. Soc.*, 94, 7040, 1972.

18 **Schmid, P., Ingold, K. U.**, Kinetic application of EPR spectroscopy. 31. Rate constants for spin trapping. 1. Primary alkyl radicals, *J. Am. Chem. Soc.*, 100, 2493, 1978.

19 **Mendenhall, G. D., Ingold K. U.**, Kinetic application of EPR spectroscopy. 9. Preparation and properties of di-*tert*-butyliminoxy radicals, *J. Am. Chem. Soc.*, 95, 2963, 1973.

20 **Calder, A., Forrester A. R., McConnachie, G.**, Nitroxide radicals. Part 14. Decomposition of 1- and 2-naphthyl *tert*-butyl nitroxides, *J. C. S. Perkin I*, 1974, 2198.

21 **Castelhano, A. L., Griller, D., Ingold, K. U.**, Do spin adducts of 5,5-dimethylpyrroline-N-oxide dimerise?, *Can. J. Chem./*, 60, 1501, 1982.

22 **Haire, D. L., Janzen, E. G.**, Synthesis and spin trapping kinetics of new alkyl substituted cyclic nitrones, *Can. J. Chem.*, 60, 1514, 1982.

23 **Gerchikov, A. Ya., Nasyrov, I. Sh., Lanina, T. P., Akmanova, N. A., Martemyanov, V. S.**, Reactivity of alkylperoxyl radicals in reaction with aromatic nitrons, *React. Kinet. Catal. Lett.*, 32, 533, 1986.

24 **Gerchikov, A. Ya., Ulanovskaya, Yu. V.**, Rate constants and pecularites of reaction of peroxyl radicals of esters with carbomaylamidonitrones, *Kinet. Katal.*, 32 , 1073, 1991.

25 **Gerchikov, A. Ya., Nasyrov, I. Sh., Akmanova, N. A.**, Reactivity of aromatic nitrones with alkyl and alkylperoxyl radicals, *React. Kinet. Catal. Lett.*, 33, 317, 1987.

Chapter 6

BOND DISSOCIATION ENERGIES AND RATE CONSTANTS OF REACTIONS OF THIOPHENOLS

Table 6.1
Dissociation energies of S—H-bonds of thiophenols[1]

No.	Name and formula of thiophenol	Formula	Mol. wt.	D_{S-H} / kJ mol^{-1}
243	1-Thionaphthol	$1\text{-}C_{10}H_7SH$	160.24	328.4
244	2-Thionaphthol	$2\text{-}C_{10}H_7SH$	160.24	326.8
245	Thiophenol	C_6H_5SH	110.18	338.1
246	Thiophenol, 3-cloro	$3\text{-}ClC_6H_4SH$	144.63	335.8
247	Thiophenol, 4-cloro	$4\text{-}ClC_6H_4SH$	144.63	334.1
248	Thiophenol, 4-ethoxy	$4\text{-}CH_3CH_2OC_6H_4SH$	154.23	329.6
249	Thiophenol, 4-methoxy	$4\text{-}CH_3OC_6H_4SH$	140.21	329.3
250	Thiophenol, 3-methyl	$3\text{-}CH_3C_6H_4SH$	124.21	338.1
251	Thiophenol, 4-methyl	$4\text{-}CH_3C_6H_4SH$	124.21	334.8
252	Thiophenol, 4-*tert*-butyl	$4\text{-}(CH_3)_3CC_6H_4SH$	166.29	337.7
253	Thiophenol, 4-trifluoromethyl	$4\text{-}CF_3C_6H_4SH$	178.18	338.6
254	Thiophenol, 2,4,6-trimethyl	$2,4,6\text{-}(CH_3)_3C_6H_2SH$	152.26	336.6

Table 6.2
Enthalpies, activation energies and rate constants of reaction of peroxyl radicals with thiophenols in hydrocarbon solutions calculated by formulas 1.15–1.17 and 1.20. The values of A, br_e and α see Table 1.6

No.	Thiophenol	$RO_2^•$	ΔH/ kJ mol^{-1}	E/ kJ mol^{-1}	k(333 K)/ l mol^{-1} s^{-1}
243	1-$C_{10}H_7SH$	$HO_2^•$	–40.6	10.9	6.2×10^5
243	1-$C_{10}H_7SH$	*sec*-$RO_2^•$	–37.1	11.9	1.3×10^5
243	1-$C_{10}H_7SH$	*tert*-$RO_2^•$	–30.2	13.9	2.1×10^5
244	2-$C_{10}H_7SH$	$HO_2^•$	–42.2	10.5	7.2×10^5
244	2-$C_{10}H_7SH$	*sec*-$RO_2^•$	–38.7	11.4	5.2×10^5
244	2-$C_{10}H_7SH$	*tert*-$RO_2^•$	–31.8	13.4	2.5×10^5
245	C_6H_5SH	$HO_2^•$	–30.9	13.7	2.3×10^5
245	C_6H_5SH	*sec*-$RO_2^•$	–27.4	14.7	1.6×10^5
245	C_6H_5SH	*tert*-$RO_2^•$	–20.5	16.9	7.1×10^4
246	3-ClC_6H_4SH	$HO_2^•$	–33.2	13.0	2.9×10^5
246	3-ClC_6H_4SH	*sec*-$RO_2^•$	–29.7	14.0	2.0×10^5
246	3-ClC_6H_4SH	*tert*-$RO_2^•$	–22.8	16.1	9.5×10^5
247	4-ClC_6H_4SH	$HO_2^•$	–34.9	12.5	3.5×10^5
247	4-ClC_6H_4SH	*sec*-$RO_2^•$	–31.4	13.5	2.4×10^5
247	4-ClC_6H_4SH	*tert*-$RO_2^•$	–24.5	15.6	1.1×10^5
248	4-$CH_3CH_2OC_6H_4SH$	$HO_2^•$	–39.4	11.3	5.4×10^5
248	4-$CH_3CH_2OC_6H_4SH$	*sec*-$RO_2^•$	–35.9	12.2	3.9×10^5
248	4-$CH_3CH_2OC_6H_4SH$	*tert*-$RO_2^•$	–29.0	14.2	1.9×10^5
249	4-$CH_3OC_6H_4SH$	$HO_2^•$	–39.7	11.2	5.6×10^5
249	4-$CH_3OC_6H_4SH$	*sec*-$RO_2^•$	–36.2	12.1	4.0×10^5
249	4-$CH_3OC_6H_4SH$	*tert*-$RO_2^•$	–29.3	14.2	1.9×10^5
250	3-$CH_3C_6H_4SH$	$HO_2^•$	–30.9	13.7	2.3×10^5
250	3-$CH_3C_6H_4SH$	*sec*-$RO_2^•$	–27.4	14.7	1.6×10^5
250	3-$CH_3C_6H_4SH$	*tert*-$RO_2^•$	–20.5	16.9	7.1×10^4
251	4-$CH_3C_6H_4SH$	$HO_2^•$	–34.2	12.7	3.3×10^5
251	4-$CH_3C_6H_4SH$	*sec*-$RO_2^•$	–30.7	13.7	2.3×10^5

No.	Thiophenol	$RO_2^•$	ΔH/ kJ mol^{-1}	E/ kJ mol^{-1}	k(333 K)/ l mol^{-1} s^{-1}
251	$4\text{-}CH_3C_6H_4SH$	*tert*-$RO_2^•$	−23.8	15.8	1.1×10^5
252	$4\text{-}(CH_3)_3CC_6H_4SH$	$HO_2^•$	−31.3	13.5	2.4×10^5
252	$4\text{-}(CH_3)_3CC_6H_4SH$	*sec*-$RO_2^•$	−27.8	14.6	1.6×10^5
252	$4\text{-}(CH_3)_3CC_6H_4SH$	*tert*-$RO_2^•$	−20.9	16.7	7.7×10^4
253	$4\text{-}F_3CC_6H_4SH$	$HO_2^•$	−30.4	13.9	2.1×10^5
253	$4\text{-}F_3CC_6H_4SH$	*sec*-$RO_2^•$	−26.9	14.9	1.5×10^5
253	$4\text{-}F_3CC_6H_4SH$	*tert*-$RO_2^•$	−20.0	17.0	6.9×10^4
254	$2,4,6\text{-}(CH_3)_3C_6H_2SH$	$HO_2^•$	−32.4	13.3	2.6×10^5
254	$2,4,6\text{-}(CH_3)_3C_6H_2SH$	*sec*-$RO_2^•$	−28.9	14.3	1.8×10^5
254	$2,4,6\text{-}(CH_3)_3C_6H_2SH$	*tert*-$RO_2^•$	−22.0	16.4	8.6×10^4

Table 6.3
Rate constants and stoichiometric coefficients *f* of reactions of cumylperoxyl radicals with phenolsulfides in cumene

No.	Phenolsulfide	k/(333 K) l mol^{-1} s^{-1}	f	Ref.
255	Allyl-(2-hydroxy-5-*tert*-butylphenyl)sulfide; $2\text{-}HO\text{-}5\text{-}(CH_3)_3CC_6H_3SCH_2CH{=}CH_2$	6.7×10^3	1.1	2
256	Benzyl-(2-hydroxy-5-*tert*-butylphenyl)sulfide; $2\text{-}HO\text{-}5\text{-}(CH_3)_3CC_6H_3SCH_2C_6H_5$	7.3×10^3	1.0	2
257	*tert*-Butyl-(2-hydroxy-5-*tert*-butylphenyl)sulfide; $2\text{-}HO\text{-}5\text{-}(CH_3)_3CC_6H_3SC(CH_3)_3$	1.0×10^3	0.9	2
258	Di-(2-hydroxy-5-*tert*-butylphenyl)sulfide; $[2\text{-}HO\text{-}5\text{-}(CH_3)_3CC_6H_3S]_2$	6.1×10^4	4.4	3
259	Di-(2-hydroxy-5-methylphenyl)sulfide; $[2\text{-}HO\text{-}5\text{-}(CH_3)_3CC_6H_3S]_2$	2.8×10^4	2.0	3
260	Dimethylaminomethylene-(2-hydroxy-3-methylthio-5-*tert*-butylphenyl); $2\text{-}HO\text{-}3\text{-}CH_3S\text{-}5\text{-}(CH_3)_3CC_6H_2SCH_2N(CH_3)_2$	5.7×10^3	1.9	4
261	Ethane-1,2-bis-(2-hydroxy-5-*tert*-butylphenyl)sulfide; $[2\text{-}HO\text{-}5\text{-}(CH_3)_3CC_6H_3SCH_2]_2$	4.6×10^4	2.0	3
262	Ethyl-(2-hydroxy-5-*tert*-butylphenyl)sulfide; $2\text{-}HO\text{-}5\text{-}(CH_3)_3CC_6H_3SCH_2CH_3$	1.8×10^4	0.8	2
263	Heptyl-(2-hydroxy-5-methylbenzyl)sulfide; $2\text{-}HO\text{-}5\text{-}CH_3C_6H_3CH_2S(CH_2)_6CH_3$	8.0×10^4	1.6	5

No.	Phenolsulfide	k/(333 K) l mol^{-1} s^{-1}	f	Ref.
264	2-Hydroxy-5-*tert*-butylphenyl sulfide; $[2-HO-5-(CH_3)_3CC_6H_3]_2S$	1.5×10^4	1.8	3
265	Isopropyl-(2-hydroxy-5-*tert*-butylphenyl)sulfide; $2-HO-5-(CH_3)_3CC_6H_3SCH(CH_3)_2$	2.0×10^4	0.9	2
266	2-Mercapto-3-N-anilinopropyl-(3,5-di-*tert*-butyl-4-hydroxyphenyl)sulfide; $4-HO-3.5-[(CH_3)_3C]_2C_6H_2SCH_2CH(SH)CH_2NHC_6H_5$	4.5×10^4		6
267	2-Mercapto-3-N-anilinopropyl-(2-hydroxy-5-*tert*-butyl)sulfide; $2-HO-5-(CH_3)_3CC_6H_3SCH_2CH(SH)CH_2NHC_6H_5$	6.0×10^3		6
268	Methylen-bis-(2-hydroxy-5-*tert*-butylphenyl)(pentyl)di sulfide; $2-HO-5-(CH_3)_3CC_6H_3SCH_2S(CH_2)_4CH_3$	6.0×10^4	1.0	5
269	Methylen-*bis*-(2-hydroxy-5-*tert*-butylphenyl) di sulfide; $[2-HO-5-(CH_3)_3CC_6H_3S]_2CH_2$	3.4×10^4	1.9	3
269	Methylen-*bis*-(2-hydroxy-5-*tert*-butylphenyl) di sulfide; $[2-HO-5-(CH_3)_3CC_6H_3S]_2CH_2$	3.9×10^4	1.9	6
270	Methyl-(2-hydroxy-5-*tert*-butyl)sulfide; $2-HO-5-(CH_3)_3CC_6H_3SCH_3$	5.8×10^4	1.2	5
271	N-Methylenmorpholyl-(2-hydroxy-3-methyl-5-*tert*-butylphenyl)sulfide; $2-HO-3-CH_3-5-(CH_3)_3CC_6H_2SCH_2$-*cyclo*-$[N(CH_2)_2O(CH_2)_2]$	3.6×10^3	1.4	4
272	N-Methylenpiperidine-(2-hydroxy-3-methylthio-5-*tert*-butylphenyl)sulfide; $2-HO-3-CH_3S-5-(CH_3)_3CC_6H_2SCH_2$-*cyclo*-$[N(CH_2)_2O(CH_2)_2]$	9.3×10^3	1.6	4
273	3-Methylthio-4-hydroxythiophenol; $3-CH_3S-4-HOC_6H_3SH$	5.6×10^4	3.4	4
274	Phenyl-(2-hydroxy-5-methylbenzyl)sulfide; $2-HO-5-C_6H_5CH_2C_6H_3CH_2SC_6H_5$	1.0×10^4	3.2	5
275	Phenyl-(2-hydroxy-3-methylthio-5-*tert*-butylphenyl)sulfide; $2-HO-3-CH_3S-5-(CH_3)_3CC_6H_2CH_2SC_6H_5$	1.0×10^4	3.2	5
275	Phenyl-(2-hydroxy-3-methylthio-5-*tert*-butylphenyl)sulfide; $2-HO-3-CH_3S-5-(CH_3)_3CC_6H_2CH_2SC_6H_5$	1.3×10^4	3.2	4
276	β-Propionitrile-(2-hydroxy-5-*tert*-butylphenyl)sulfide; $2-HO-5-(CH_3)_3CC_6H_3SCH_2CH_2CN$	6.0×10^3	1.1	2

Table 6.4
Enthalpies, activation energies and rate constants of reaction of $C_6H_5S^•$ with hydrocarbons (R_3H) and hydroperoxides in hydrocarbon solutions calculated by formulas 1.15–1.17 and 1.20. The values of A, br_e and α, see Table 1.6

R_3H (ROOH)	ΔH/ kJ mol^{-1}	E/ kJ mol^{-1}	k(333 K)/ l mol^{-1} s^{-1}	k(400 K)/ l mol^{-1} s^{-1}
$C_6H_5CH_3$	30.1	46.6	14.7	2.5×10^2
4-$CH_3C_6H_4CH_3$	27.3	44.8	56.3	1.3×10^3
4-$CH_3OC_6H_4CH_3$	27.1	44.7	29.2	8.7×10^2
4-$ClC_6H_4CH_3$	29.2	46.1	17.6	2.9×10^2
4-$CNC_6H_4CH_3$	23.8	42.6	62.3	8.2×10^2
4-$NO_2C_6H_4CH_3$	31.4	47.5	10.6	2.1×10^2
$C_6H_5CH_2CH_3$	18.9	39.6	1.2×10^3	1.3×10^3
$C_6H_5CH(CH_3)_2$	12.5	35.7	2.5×10^2	2.2×10^3
$(C_6H_5)_2CH_2$	11.9	35.3	5.8×10^2	4.9×10^3
$(C_6H_5)_2CHCH_3$	6.1	31.9	9.9×10^2	6.8×10^3
$C_6H_5CH_2CH_2C_6H_5$	18.1	39.1	2.9×10^2	3.1×10^3
$(C_6H_5)_3CH$	4.7	31.1	1.3×10^3	8.7×10^3
cyclo-$C_6H_{11}C_6H_5$	11.5	35.1	3.1×10^2	2.6×10^3
Indane	16.7	38.2	4.1×10^2	4.1×10^3
Tetraline	10.9	34.7	1.4×10^3	1.2×10^4
1,2-Dihydronaphthaline	−8.4	23.8	7.4×10^4	3.2×10^5
1,2-Dihydropyrene	−3.5	26.5	2.8×10^4	1.4×10^5
9,10-Dihydrophenanthrane	−1.2	27.7	1.8×10^4	9.7×10^4
H_2O_2	30.9	44.7	6.2	93.1
sec-ROOH	27.4	42.2	7.7	99
tert-ROOH	20.5	37.5	41.9	4.1×10^2

Table 6.5
Enthalpies, activation energies and rate constants of reaction of *para*-metoxyphenoxyl radical with thiophenols in hydrocarbon solutions calculated by formulas 1.15–1.17 and 1.20. The values of A, br_e and α, see Table 1.6

No.	Thiophenol	ΔH/ kJ mol^{-1}	E/ kJ mol^{-1}	k(333 K)/ l mol^{-1} s^{-1}	k(400 K)/ l mol^{-1} s^{-1}
243	1-$C_{10}H_7SH$	–20.6	19.0	1.0×10^5	3.3×10^5
244	2-$C_{10}H_7SH$	–22.2	18.5	1.3×10^5	3.8×10^5
245	C_6H_5SH	–10.9	22.2	3.3×10^4	1.3×10^5
246	3-ClC_6H_4SH	–13.2	21.4	4.4×10^4	1.6×10^5
247	4-ClC_6H_4SH	–14.9	20.9	5.3×10^4	1.9×10^5
248	4-$CH_3CH_2OC_6H_4SH$	–19.4	19.4	9.1×10^4	2.9×10^5
249	4-$CH_3OC_6H_4SH$	–19.7	19.3	9.4×10^4	3.0×10^5
250	3-$CH_3C_6H_4SH$	–10.9	22.2	3.3×10^4	1.3×10^5
251	4-$CH_3C_6H_4SH$	–14.2	21.1	4.9×10^4	1.8×10^5
252	4-$(CH_3)_3CC_6H_4SH$	–11.3	22.1	3.4×10^4	1.3×10^5
253	4-$CF_3C_6H_4SH$	–10.4	22.4	3.1×10^4	1.2×10^5
254	2,4,6-$(CH_3)_3C_6H_2SH$	–12.4	21.7	3.9×10^4	1.5×10^5

Table 6.6
Enthalpies, activation energies and rate constants of reaction of $C_6H_5S^\bullet$ with phenols in nonpolar solutions calculated by formulas 1.15–1.17 and 1.20. The values of A, br_e and α, see Table 1.6

No.	Phenol	ΔH/ kJ mol^{-1}	E/ kJ mol^{-1}	k(333 K)/ l mol^{-1} s^{-1}	k(400 K)/ l mol^{-1} s^{-1}
3	2-HOC_6H_4OH	–0.1	24.6	2.8×10^4	1.2×10^5
4	2-HO-4-$(CH_3)_3CC_6H_3OH$	4.3	27.4	1.0×10^4	5.2×10^4
9	4-$CH_3C_6H_4OH$	19.7	37.6	1.3×10^2	1.2×10^3
10	$(CH_3)_3C_6(OH)(CH_2)_3O$	–9.3	18.9	1.1×10^5	3.4×10^5
20	$(CH_3)_3C_6(OH)CH_2CH(CH_3)O$	–13.7	16.3	2.6×10^5	7.4×10^5
25	1-$HOC_{13}H_9$	–1.8	23.6	2.0×10^4	8.3×10^4
29	1-$HOC_{10}H_7$	3.3	26.8	6.2×10^3	3.2×10^4

No.	Phenol	ΔH/ kJ mol^{-1}	E/ kJ mol^{-1}	k(333 K)/ l mol^{-1} s^{-1}	k(400 K)/ l mol^{-1} s^{-1}
30	$2\text{-}HOC_{10}H_7$	13.7	33.6	5.4×10^2	4.1×10^3
32	$1\text{-}HOC_{14}H_9$	14.6	34.2	4.3×10^2	3.4×10^3
36	C_6H_5OH	28.9	44.0	12.5	1.8×10^2
38	$4\text{-}CH_3COC_6H_4OH$	31.4	45.8	6.5	1.0×10^2
40	$4\text{-}NH_2C_6H_4OH$	16.6	35.5	2.7×10^2	2.3×10^3
41	$4\text{-}C_6H_5CH_2OC_6H_4OH$	9.4	30.7	1.5×10^3	9.8×10^3
44	$4\text{-}CH_3(CH_2)_3OC_6H_4OH$	7.7	29.6	2.3×10^3	1.4×10^4
46	$4\text{-}(CH_3)_3CC_6H_4OH$	18.8	37.0	1.6×10^2	1.5×10^3
47	$2,4\text{-}[(CH_3)_3C]_2C_6H_3OH$	19.4	37.4	1.4×10^2	1.3×10^3
74	$2\text{-}(CH_3)_3C\text{-}4,6\text{-}(CH_3)_2C_6H_2OH$	15.7	34.9	3.4×10^2	2.8×10^3
75	$2\text{-}(CH_3)_3C\text{-}4\text{-}CH_3OC_6H_3OH$	3.3	26.8	6.2×10^3	3.2×10^4
78	$4\text{-}CH_3OC_6H_4OH$	10.9	31.7	1.1×10^3	7.2×10^3
79	$2,3\text{-}(CH_3)_2C_6H_3OH$	15.4	34.7	3.6×10^2	2.9×10^3
80	$3,4\text{-}(CH_3)_2C_6H_3OH$	14.5	34.1	4.5×10^2	3.5×10^3
81	$3,5\text{-}(CH_3)_2C_6H_3OH$	15.9	35.1	3.1×10^2	2.6×10^3
82	$2,4\text{-}(CH_3)_2C_6H_3OH$	20.4	38.1	1.1×10^2	1.1×10^3
83	$2,6\text{-}(CH_3)_2C_6H_3OH$	16.8	35.7	2.5×10^2	2.2×10^3
85	$4\text{-}(CH_3)_2NC_6H_4OH$	−11.2	17.8	1.6×10^5	4.7×10^5
110	$2,4,6\text{-}(CH_3)_3C_6H_2OH$	−6.6	20.6	5.9×10^4	2.0×10^5
130	$\alpha\text{-}(CH_3)_3C_6(OH)(CH_2)_2OC(CH_3)[(CH_2)_3CH(CH_3)]_3CH_3$	−10.4	18.3	1.3×10^5	4.1×10^5
131	$\beta\text{-}(CH_3)_2C_6H(OH)(CH_2)_2OC(CH_3)[(CH_2)_3CH(CH_3)]_3CH_3$	−5.7	21.1	4.9×10^4	1.8×10^5
133	$\delta\text{-}CH_3C_6H_2(OH)(CH_2)_2OC(CH_3)[(CH_2)_3CH(CH_3)]_3CH_3$	−21.1	12.0	1.3×10^6	2.7×10^6

Table 6.7
Enthalpies, activation energies and rate constants of reaction of nitroxyl radicals with thiophenols calculated by formulas 1.15–1.17 and 1.20. The values of A, br_e and α, see Table 1.6

No.	Thiophenol	ΔH/ kJ mol^{-1}	E/ kJ mol^{-1}	k(333 K)/ l mol^{-1} s^{-1}	k(400 K)/ l mol^{-1} s^{-1}
	$[(CH_3)_3C]_2C{=}NO^{\bullet}$				
243	$1\text{-}C_{10}H_7SH$	−10.1	29.9	2.0×10^3	1.2×10^4
244	$2\text{-}C_{10}H_7SH$	−11.7	29.3	2.5×10^3	1.5×10^4
245	C_6H_5SH	−0.4	33.4	5.6×10^2	4.3×10^3
246	$3\text{-}ClC_6H_4SH$	−2.7	32.6	7.7×10^2	5.5×10^3
247	$4\text{-}ClC_6H_4SH$	−4.4	31.9	9.9×10^2	6.8×10^3
248	$4\text{-}CH_3CH_2OC_6H_4SH$	−8.9	30.3	1.8×10^3	1.1×10^4
249	$4\text{-}CH_3OC_6H_4SH$	−9.2	30.2	1.8×10^3	1.1×10^4
251	$4\text{-}CH_3C_6H_4SH$	−3.7	32.2	8.9×10^2	6.2×10^3
253	$4\text{-}CF_3C_6H_4SH$	0.1	33.6	5.4×10^2	4.1×10^3
254	$2,4,6\text{-}(CH_3)_3C_6H_2SH$	−1.9	32.8	7.2×10^2	5.2×10^2
	cyclo-$[C(CH_3)_2N{=}C(C_6H_5)N(O^{\bullet})]$				
243	$1\text{-}C_{10}H_7SH$	16.2	39.9	57.1	6.2×10^2
244	$2\text{-}C_{10}H_7SH$	14.6	39.2	70.9	7.6×10^2
245	C_6H_5SH	25.9	44.0	12.5	1.8×10^2
246	$3\text{-}ClC_6H_4SH$	23.6	43.0	18.0	2.4×10^2
247	$4\text{-}ClC_6H_4SH$	21.9	42.3	23.2	3.0×10^2
248	$4\text{-}CH_3CH_2OC_6H_4SH$	17.4	40.4	46.0	5.3×10^2
249	$4\text{-}CH_3OC_6H_4SH$	17.1	40.3	47.7	5.5×10^2
251	$4\text{-}CH_3C_6H_4SH$	22.6	42.6	20.8	2.7×10^2
253	$4\text{-}CF_3C_6H_4SH$	26.4	44.2	11.6	1.7×10^2
254	$2,4,6\text{-}(CH_3)_3C_6H_2SH$	24.4	43.4	15.6	2.1×10^2

No.	Thiophenol	ΔH/ kJ mol^{-1}	E/ kJ mol^{-1}	k(333 K)/ l mol^{-1} s^{-1}	k(400 K)/ l mol^{-1} s^{-1}
	cyclo-[$C(CH_3)_2CH_2COCH_2C(CH_3)_2N(O^{\bullet})$]				
243	$1\text{-}C_{10}H_7SH$	28.0	44.9	9.1	1.4×10^2
244	$2\text{-}C_{10}H_7SH$	26.4	44.2	11.7	1.7×10^2
245	C_6H_5SH	37.7	49.4	1.8	35.4
246	$3\text{-}ClC_6H_4SH$	35.4	48.3	2.6	49.2
247	$4\text{-}ClC_6H_4SH$	33.7	47.5	3.5	62.6
248	$4\text{-}CH_3CH_2OC_6H_4SH$	29.2	45.5	7.3	1.1×10^2
249	$4\text{-}CH_3OC_6H_4SH$	28.9	45.3	7.8	1.2×10^2
251	$4\text{-}CH_3C_6H_4SH$	34.4	47.8	3.2	57.2
253	$4\text{-}CF_3C_6H_4SH$	38.6	49.6	1.7	33.3
254	$2,4,6\text{-}(CH_3)_3C_6H_2SH$	36.2	48.7	2.3	43.6

Table 6.8
Activation energies and rate constants of reaction $ArSH + O_2 \rightarrow ArS^{\bullet} + HO_2^{\bullet}$ calculated by formula 1.17 with $A = 10^{10}$ l mol^{-1} s^{-1}

No.	Thiophenol	E/ kJ mol^{-1}	k(400 K)/ l mol^{-1} s^{-1}	k(500 K)/ l mol^{-1} s^{-1}
243	$1\text{-}C_{10}H_7SH$	125.0	4.7×10^{-7}	8.7×10^{-4}
244	$2\text{-}C_{10}H_7SH$	123.4	8.7×10^{-7}	1.3×10^{-3}
245	C_6H_5SH	134.7	2.6×10^{-8}	8.5×10^{-5}
246	$3\text{-}ClC_6H_4SH$	132.4	5.1×10^{-8}	1.5×10^{-4}
247	$4\text{-}ClC_6H_4SH$	130.7	8.5×10^{-8}	2.2×10^{-4}
248	$4\text{-}CH_3CH_2OC_6H_4SH$	126.2	3.3×10^{-7}	6.5×10^{-4}
249	$4\text{-}CH_3OC_6H_4SH$	125.6	3.6×10^{-7}	7.0×10^{-4}
250	$3\text{-}CH_3C_6H_4SH$	134.7	2.6×10^{-8}	8.5×10^{-5}
251	$4\text{-}CH_3C_6H_4SH$	131.4	6.9×10^{-8}	1.9×10^{-4}
252	$4\text{-}(CH_3)_3CC_6H_4SH$	134.3	2.9×10^{-8}	9.3×10^{-5}
253	$4\text{-}CF_3C_6H_4SH$	135.2	2.2×10^{-8}	7.5×10^{-5}
254	$2,4,6\text{-}(CH_3)_3C_6H_2SH$	133.2	4.0×10^{-8}	1.2×10^{-4}

REFERENCES

1 **Denisov, E. T.**, Estimation of S—H-bond dissociation energies in thiophenols and thioalcohols from kinetic data, *Zh. Fiz. Khim.*, (in press).

2 **Kuliev, F. A.**, *The study of physico-chemical properties and reactivity of oxyphenilfides as inhibitors of oxidation*, Cand. Sci. (Chem.) Thesis Dissertation, Inst. Chem. Phys., Chernogolovka, 1979 (in Russian).

3 **Akhundova, M. M.**, *The mechanism of inhibiting action of some thiobisphenols and aminesulfides in oxidizing hydrocarbons*, Cand. Sci. (Chem.) Thesis Dissertation, Inst. Chem. Phys., Chernogolovka, 1982 (in Russian).

4 **Mamedov, F. A.**, *Substituted phenols and mercaptophenols as inhibitors of metal corrosion and additives to lubricants*, Doct. Sci. (Chem.) Thesis Dissertation, Baku, 1991 (in Russian).

5 **Aliev, A. S.**, *The mechanism of antioxidative action of some phenolic compounds*, Cand. Sci. (Chem.) Thesis Dissertation, Inst. Chem. Phys., Chernogolovka, 1975 (in Russian).

6 **Aliev, A. S., Farzaliev, V. M., Denisov, E. T.**, The mechanism of antioxidative action of oxyphenylsulfides on cumene oxidation, *Neftekhimiya*, 15, 890, 1975.

Chapter 7

PHOSPHORUS AND SULFUR CONTAINING ANTIOXIDANTS DECOMPOSING HYDROPEROXIDES

Table 7.1
Rate constants of reaction of peroxyl radicals with phosphines and phosphites in hydrocarbon solutions

No.	Phosphite or phosphine	$RO_2^{\bullet}$	k(338 K)/ $l\ mol^{-1}\ s^{-1}$ or $\log k = A - E/\theta$	Ref.
277	2,2'-Biphenylenisopropylphosphite; $C_{12}H_8O_2POCH(CH_3)_2$	$(CH_3)(CN)CO_2^{\bullet}$	680	1
278	2-*tert*-Butyl-4-methylphenylphosphite; $(2\text{-}(CH_3)_3C\text{-}4\text{-}CH_3C_6H_3O)_3P$	$C_6H_5(CH_3)_2CO_2^{\bullet}$	80	1
279	Chlorodiphenylphosphine; $(C_6H_5)PCl$	$(CH_3)_3CO_2^{\bullet}$	3.7 – 5.4/θ	2
280	2,6-Di-*tert*-butyl-4-methylphenyl-1,3-(2,2-dimethyl)propylenphosphite; *cyclo*-$[OCH_2C(CH_3)_2CH_2O]POC_6H_2\text{-}2,6\text{-}[(CH_3)_3C]_2\text{-}4\text{-}CH_3$	$C_6H_5(CH_3)_2CO_2^{\bullet}$	150	1
281	Diphenylphosphine; $(C_6H_5)_2PH$	$(CH_3)_3CO_2^{\bullet}$	5.0 – 10.4/θ	2
282	2,2'-Methylen-bis-(4-*tert*-butyl-6-methylphenyl)-2,6-di-*tert*-butyl-4-methylphenylphosphite; $CH_2\text{-}2,2'\text{-}[3\text{-}CH_3\text{-}5\text{-}(CH_3)_3CC_6H_2O]_2POC_6H_2\text{-}2,6\text{-}[(CH_3)_3C]_2\text{-}4\text{-}CH_3$	$C_6H_5(CH_3)_2CO_2^{\bullet}$	40	1
283	2,2'-Methylen-bis-(4-*tert*-butyl-6-methylphenyl)-phenylphosphite; $CH_2\text{-}2,2'\text{-}[3\text{-}CH_3\text{-}5\text{-}(CH_3)_3CC_6H_2O]_2POC_6H_5$	$C_6H_5(CH_3)_2CO_2^{\bullet}$	50	1
284	Methylethylenglicylphosphite; *cyclo*-$[CH_2OP(OCH_3)OCH_2]$	$(CH_3)_3CO_2^{\bullet}$	6.0 – 19.2/θ	2
285	Methoxydiphenylphosphine; $(C_6H_5)_2POCH_3$	$(CH_3)_3CO_2^{\bullet}$	3.7 – 5.4/θ	2
286	Methyldiphenylphosphine; $(C_6H_5)_2PCH_3$	$(CH_3)_3CO_2^{\bullet}$	5.1 – 6.3/θ	2

No.	Phosphite or phosphine	$RO_2^{\bullet}$	k(338 K)/ l mol^{-1} s^{-1} or log $k = A - E/\theta$	Ref.
287	1,2-Phenylen-2,6-di-*tert*-butyl-4-methylphenylphosphite; $1,2\text{-}C_6H_4O_2PO\text{-}C_6H_2\text{-}2,6\text{-}[(CH_3)_3C]_2\text{-}4\text{-}CH_3$	$C_6H_5(CH_3)_2CO_2^{\bullet}$	140	1
288	1,2-Phenylenisopropylphosphite; $1,2\text{-}C_6H_4O_2POCH(CH_3)_2$	$(CH_3)_2(CN)CO_2^{\bullet}$	940	1
289	Triallylphosphite; $(CH_2{=}CHCH_2O)_3P$	$(CH_3)_3CO_2^{\bullet}$	5.7 – 15.5/θ	2
290	Tri-*tert*-butylphosphite; $[(CH_3)_3CO]_3P$	$(CH_3)_3CO_2^{\bullet}$	6.3 – 23.0/θ	2
290	Tri-*tert*-butylphosphite; $[(CH_3)_3CO]_3P$	$(CH_3)_2(CN)CO_2^{\bullet}$	740	1
291	Tri-(*p*-chlorophenyl)phosphine; $(4\text{-}ClC_6H_4)_3P$	$(CH_3)_3CO_2^{\bullet}$	6.8 – 17.1/θ	2
292	Triethylphosphite; $(CH_3CH_2O)_3P$	$(CH_3)_3CO_2^{\bullet}$	5.0 – 13.8/θ	2
292	Triethylphosphite; $(CH_3CH_2O)_3P$	$(CH_3)_2(CN)CO_2^{\bullet}$	1900	1
293	Tri-(*p*-fluorophenyl)phosphine; $(4\text{-}FC_6H_4)_3P$	$(CH_3)_3CO_2^{\bullet}$	5.9 – 13.0/θ	2
294	Tri-*iso*-propylphosphite; $[(CH_3)_2CH]_3P$	$(CH_3)_3CO_2^{\bullet}$	6.2 – 18.0/θ	2
294	Tri-*iso*-propylphosphite; $[(CH_3)_2CH]_3P$	$(CH_3)_2(CN)CO_2^{\bullet}$	1300	1
295	Tri-(*p*-methoxyphenyl)phosphine; $(4\text{-}CH_3OC_6H_4)_3P$	$(CH_3)_3CO_2^{\bullet}$	5.9 – 13.4/θ	2
296	Tri-(*p*-methylphenyl)phosphine; $(4\text{-}CH_3C_6H_4)_3P$	$(CH_3)_3CO_2^{\bullet}$	6.3 – 15.9/θ	2
297	Trimethylphosphite; $(CH_3O)_3P$	$(CH_3)_3CO_2^{\bullet}$	5.7 – 15.9/θ	2
298	Triphenylphosphine; $(C_6H_5)_3P$	$(CH_3)_3CO_2^{\bullet}$	6.0 – 12.5/θ	2
299	Triphenylphosphite; $(C_6H_5O)_3P$	$(CH_3)_3CO_2^{\bullet}$	7.3 – 20.1/θ	2

Table 7.2
Rate constants of reaction of hydroperoxide with phosphites

No.	Phosphite (phosphine)	ROOH	Solvent	k(333 K)/ l mol^{-1} s^{-1} or log k = $A - E/\theta$	Ref.
280	*cyclo*-[$OCH_2C(CH_3)_2CH_2OP(O$-2,6-$[(CH_3)_3C]_2$-4-$CH_3C_6H_2)$]	$C_6H_5C(CH_3)_2OOH$	C_6H_5Cl	$7.88 - 54.0/\theta$	3
282	CH_2-[3-CH_3-5-$(CH_3)_3CC_6H_3$-6-O$]_2$PO-2,6-$[(CH_3)_3C]_2$-4-$CH_3C_6H_2$	$C_6H_5C(CH_3)_2OOH$	C_6H_5Cl	$10.94 - 77.0/\theta$	3
287	1,2-$C_6H_4O_2$PO-2,6-$[(CH_3)_3C]_2$-4-$CH_3C_6H_2$	$C_6H_5C(CH_3)_2OOH$	C_6H_6	$5.50 - 45.6/\theta$	4
287	1,2-$C_6H_4O_2$PO-2,6-$[(CH_3)_3C]_2$-4-$CH_3C_6H_2$	$C_6H_5C(CH_3)_2OOH$	C_6H_5Cl	$10.0 - 72.0/\theta$	5
288	1,2-$C_6H_4O_2POCH(CH_3)_2$	$C_6H_5C(CH_3)_2OOH$	C_6H_5Cl	0.19	3
298	$(C_6H_5)_3P$	*prim*-C_4H_9OOH	C_6H_{14}	$4.9 - 19.7/\theta$	6
298	$(C_6H_5)_3P$	*prim*-C_4H_9OOH	CH_2Cl_2	$7.4 - 35.1/\theta$	6
298	$(C_6H_5)_3P$	*prim*-C_4H_9OOH	C_2H_5OH	$7.2 - 35.1/\theta$	6
298	$(C_6H_5)_3P$	*sec*-C_4H_9OOH	C_6H_{14}	$5.8 - 27.2/\theta$	6
298	$(C_6H_5)_3P$	*sec*-C_4H_9OOH	CH_2Cl_2	$8.3 - 42.3/\theta$	6
298	$(C_6H_5)_3P$	*sec*-C_4H_9OOH	C_2H_5OH	$9.0 - 45.2/\theta$	6
298	$(C_6H_5)_3P$	*tert*-C_4H_9OOH	C_6H_{14}	$5.3 - 25.1/\theta$	6
298	$(C_6H_5)_3P$	*tert*-C_4H_9OOH	CH_2Cl_2	$7.8 - 41.0/\theta$	6
298	$(C_6H_5)_3P$	*tert*-C_4H_9OOH	C_2H_5OH	$8.8 - 46.9/\theta$	6
300	2,2′-Biphenyl-(2,6-di-*tert*-butyl-4-methyl)phenylphosphite; 2,2′-$(C_6H_4O)_2$PO-2,6-$[(CH_3)_3C]$-4-$CH_3C_6H_2$	$C_6H_5C(CH_3)_2OOH$	C_6H_5Cl	$11.62 - 77.0/\theta$	3
301	1,8-Dinaphthyl-(2,6-di-*tert*-butyl-4-methyl)phenylphoshite; 1,8-$C_{10}H_6O_2$PO-2,6-$[(CH_3)_3C]_2$-4-$CH_3C_6H_2$	$C_6H_5C(CH_3)_2OOH$	C_6H_5Cl	$9.73 - 67.0/\theta$	3
302	2,2′-Methylen-bis-(4-methyl)phenylphosphite; CH_2-[3-CH_3-C_6H_3-6-O$]_2$PO-4-$(CH_3)_3C$	$C_6H_5C(CH_3)_2OOH$	C_6H_6	$7.58 - 57.7/\theta$	4
303	1,2-Phenylenphenylphosphite; 1,2-$C_6H_4O_2POC_6H_5$	$C_6H_5C(CH_3)_2OOH$	C_6H_5Cl	990	3
304	Tri-butylphosphite; $(C_4H_9O)_3P$	$C_6H_5C(CH_3)_2OOH$	C_6H_6	$3.50 - 24.7/\theta$	4
305	Tricyclobutylphosphite; $(C_6H_{11}O)_3P$	$C_6H_5C(CH_3)_2OOH$	C_6H_6	$3.58 - 25.5/\theta$	4
306	Tri-α-naphthylphosphite; $(1\text{-}C_{10}H_7O)_3P$	$C_6H_5C(CH_3)_2OOH$	C_6H_6	$4.69 - 37.6/\theta$	4

Table 7.3
Rate constants of reaction of peroxyl radicals with sulfides and disulfides in hydrocarbon solutions

No.	Sulfide or disulfide	$RO_2^•$	T/K	k/ l mol^{-1} s^{-1}	Ref.
307	Benzylbenzylcarboxymethyldisulfide; $C_6H_5CH_2SSCH_2C(O)OCH_2C_6H_5$	$C_6H_5(CH_3)_2CO_2^•$	333	21.0	7
308	Benzyl-*tert*-butylsulfide; $C_6H_5CH_2SC(CH_3)_3$	$(CH_3)_3CO_2^•$	303	1.0×10^{-2}	8
308	Benzyl-*tert*-butylsulfide; $C_6H_5CH_2SC(CH_3)_3$	$C_6H_5CH(O_2^•)SC(CH_3)_3$	303	1.0×10^{-2}	8
309	Benzylcarboxymethyldisulfide; $C_6H_5CH_2SSCH_2C(O)OH$	$C_6H_5(CH_3)_2CO_2^•$	333	50.4	7
310	Benzylphenylsulfide; $C_6H_5CH_2SC_6H_5$	$(CH_3)_3CO_2^•$	303	0.16	8
311	Dibenzyldisulfide; $(C_6H_5CH_2S)_2$	$(CH_3)_3CO_2^•$	303	0.15	8
311	Dibenzyldisulfide; $(C_6H_5CH_2S)_2$	$C_6H_5(CH_3)_2CO_2^•$	333	7.8	7
312	Dibenzylsulfide; $(C_6H_5CH_2)_2S$	$(CH_3)_3CO_2^•$	303	0.36	8
312	Dibenzylsulfide; $(C_6H_5CH_2)_2S$	$C_6H_5CH(O_2^•)SCH_2C_6H_5$	303	0.35	8
313	Dibenzylthiosulfinate; $C_6H_5CH_2SS(O)CH_2C_6H_5$	$C_6H_5(CH_3)_2C(O_2^•)$	333	1.7×10^5	7
314	Dibenzylthiosulfonate; $C_6H_5CH_2SS(O_2)CH_2C_6H_5$	$C_6H_5(CH_3)_2C(O_2^•)$	333	8.0×10^2	7
315	Diethyldisulfide; $(CH_3CH_2S)_2$	$C_6H_5(CH_3)_2C(O_2^•)$	333	0.35	7
316	Di-2-(tridecanoyl)ethylsulfide; $C_{12}H_{25}OC(O)CH_2CH_2]_2S$	$C_6H_5(CH_3)_2CO_2^•$	333	40	9
317	Tetrahydrothiophen; *cyclo*-$[(CH_2)_4S]$	$(CH_3)_3CO_2^•$	303	8.0×10^{-2}	8

No.	Sulfide or disulfide	$RO_2^{\bullet}$	T/K	k/ l mol^{-1} s^{-1}	Ref.
317	Tetrahydrothiophen; *cyclo*-$[(CH_2)_4S]$	*cyclo*-$[CH(O_2^{\bullet})(CH_2)_3S]$	303	0.10	8
318	Tetrahydrothiopyran; *cyclo*-$[(CH_2)_5S]$	*cyclo*-$[CH(O_2^{\bullet})(CH_2)_4S]$	303	2.0×10^{-2}	8
319	*tert*-Butylsulfonic acid; $(CH_3)_3CS(O)H$	$1\text{-}C_{10}H_{11}O_2^{\bullet}$	333	1.0×10^{7}	10

Table 7.4
The rate constant of reaction of hydroperoxides with sulfides, disulfides and sulfoxides

No.	R_1SR_2	ROOH	Solvent	T/ K	k/ l mol^{-1} s^{-1} or log k = A − E/θ	Ref.
244	β-Thionaphthol; $2\text{-}C_{10}H_7SH$	$C_6H_5C(CH_3)_2OOH$	Mineral oil	423	1.0	9
307	Benzylbenzylcarboxymethyldisulfide; $C_6H_5CH_2SSCH_2COOCH_2C_6H_5$	$(CH_3)_3COOH$	C_6H_5Cl	313–373	3.0 – 21.2/θ	11
309	Benzylcarboxymethyldisulfide; $C_6H_5CH_2SSCH_2COOH$	$(CH_3)_3COOH$	C_6H_5Cl	353–373	8.1 – 59.5/θ	11
312	Dibenzylsulfide; $(C_6H_5CH_2)_2S$	$C_6H_5C(CH_3)_2OOH$	C_6H_5Cl	353	1.3×10^{-3}	12
320	Dibenzoyldisulfide; $[C_6H_5C(O)S]_2$	$C_6H_5C(CH_3)_2OOH$	C_6H_5Cl	353	1.4×10^{-2}	12
321	Dibenzoylsulfide; $[C_6H_5C(O)]_2S$	$C_6H_5C(CH_3)_2OOH$	C_6H_5Cl	353	2.3×10^{-2}	12
322	Dibutyldisulfide; $[CH_3(CH_2)_3S]_2$	$C_6H_5C(CH_3)_2OOH$	Mineral oil	423	1.0	9
323	Dibutylsulfide; $[CH_3(CH_2)_3]_2S$	$C_6H_5C(CH_3)_2OOH$	Mineral oil	423	0.67	9
324	Dibutylsulfoxide; $[CH_3(CH_2)_3]_2SO$	$C_6H_5C(CH_3)_2OOH$	Mineral oil	423	3.3×10^{-2}	9

No.	R_1SR_2	ROOH	Solvent	*T*/ K	*k*/ l mol^{-1} s^{-1} or log *k* = *A* − *E*/θ	Ref.
325	Didecylsulfide; $(C_{10}H_{21})_2S$	$C_6H_5C(CH_3)_2OOH$	Mineral oil	423	8.3	9
326	Diethyldisulfoxide; $[CH_3CH_2S(O)]_2$	$(CH_3)_3COOH$	C_6H_5Cl	373	0.39	11
327	Di-(4-hydroxy)phenylsulfide; $(4\text{-}HOC_6H_4)_2S$	$C_6H_5C(CH_3)_2OOH$	Mineral oil	423	1.0	9
328	Dimethylthiuramdisulfide; $[(CH_3)_2NC(S)S]_2$	$C_6H_5C(CH_3)_2OOH$	C_6H_5Cl	353	2.0×10^{-2}	12
329	Dimethylthiuramsulfide; $[(CH_3)_2NC(S)]_2S$	$C_6H_5C(CH_3)_2OOH$	C_6H_5Cl	353	6.2×10^{-2}	12
330	Dipropyldisulfide; $(CH_3CH_2CH_2S)_2$	$(CH_3)_3COOH$	C_6H_5Cl	353–383	11.2 − 82/θ	11
331	Diphenylsulfide; $(C_6H_5)_2S$	$C_6H_5C(CH_3)_2OOH$	Mineral oil	423	1.0×10^{-2}	9
332	Phenyl-4-hydroxyphenylsulfide; $4\text{-}HOC_6H_4SC_6H_5$	$C_6H_5C(CH_3)_2OOH$	Mineral oil	423	0.10	9

Table 7.5
Rate constants of bimolecular catalytic decomposition of hydroperoxides under action of products of oxidation of S-containing compounds in chlorobenzene

No.	S-Compound	Hydroperoxide	T/K	k/l mol^{-1} s^{-1} or log k = $A - E/\theta$	Ref.
311	Dibenzyl disulfide; $C_6H_5CH_2SSCH_2C_6H_5$	$(CH_3)_3COOH$	353–383	11.23 – 82.0/θ	7
311	Dibenzyl disulfide; $C_6H_5CH_2SSCH_2C_6H_5$	$C_6H_5(CH_3)_2COOH$	363–383	7.23 – 48.0/θ	7
312	Benzylthiuram-N-propyl; $C_6H_5CH_2CSNHC_3H_7$	$C_6H_5(CH_3)_2COOH$	343–383	9.30 – 63.0/θ	13
313	Dibenzylthio sulfinate; $C_6H_5CH_2SSOCH_2C_6H_5$	$(CH_3)_3COOH$	353–383	11.38 – 82.0/θ	7
315	Diethyl disulfide; $C_2H_5SSC_2H_5$	$C_6H_5(CH_3)_2COOH$	363–383	10.90 – 77.0/θ	7
333	Di-*tert*-butylthio sulfinate; $(CH_3)_3CSSOC(CH_3)_3$	$C_6H_5(CH_3)_2COOH$	383–403	13.59 – 106.0/θ	11
334	Benzylcarboxymethyl disulfide; $C_6H_5CH_2SSCH_2COOH$	$(CH_3)_3COOH$	353–373	8.11 – 59.5/θ	14
335	Dibenzyl sulfone; $C_6H_5CH_2SO_2CH_2C_6H_5$	$(CH_3)_3COOH$	353–383	9.76 – 76.51/θ	7
336	3-Butoxy-2-butylfhiopropyl aniline; $C_4H_9OCH_2CH(SC_4H_9)CH_2NHC_6H_5$	$C_6H_5(CH_3)_2COOH$	354–383	12.95 – 96.0/θ	15
337	N-(3-Butoxy-2-mercapto)propyl aniline; $C_4H_9OCH_2CSHHCH_2NHC_6H_5$	$C_6H_5(CH_3)_2COOH$	353–383	12.96 – 89.5/θ	15
338	Di-(*tert*-butyl-4-hydroxy)phenyl disulfide; $[3,5\text{-}[(CH_3)_3C]_2\text{-}4\text{-}HOC_6H_2]_2S_2$	$C_6H_5(CH_3)_2COOH$	393	0.51	11
339	Di-3,5-(*tert*-butyl-4-hydroxy)phenyl sulfide; $[3,5\text{-}[(CH_3)_3C]_2\text{-}4\text{-}HOC_6H_2]_2S$	$C_6H_5(CH_3)_2COOH$	393	0.38	11
340	Di-(3,5-*tert*-butyl-4-hydroxy)phenyl sulfonate; $[3,5\text{-}[(CH_3)_3C]\text{-}4\text{-}HOC_6H_2]_2SSO_2$	$C_6H_5(CH_3)_2COOH$	—	6.40 – 50.3/θ	11

No.	S-Compound	Hydroperoxide	T/ K	k / l mol^{-1} s^{-1} or log k = $A - E/\theta$	Ref.
341	Di-3,5-(*tert*-butyl-4-hydroxy)phenyl sulfoxide; $[3,5\text{-}[(CH_3)_3C]_2\text{-}4\text{-}HOC_6H_2]_2SO$	$C_6H_5(CH_3)_2COOH$	—	8.38 – 67.3/θ	11
342	Di-*tert*-butyl sulfide; $(CH_3)_3CSC(CH_3)_3$	$C_6H_5(CH_3)_2COOH$	393	0.34	11
343	Di-*tert*-butyl sulfoxide; $(CH_3)_3CSOC(CH_3)_3$	$C_6H_5(CH_3)_2COOH$	383–403	12.79 – 100.0/θ	11
344	Di-(2-hydroxy-4-*tert*-butyl)phenylmethylen disulfides; $2\text{-}HO\text{-}4\text{-}(CH_3)_3CC_6H_3)_2SCH_2S$	$C_6H_5(CH_3)_2COOH$	353–383	9.85 – 64.3/θ	16
345	Di-(2-hydroxy-5-*tert*-butyl)phenyl sulfide; $(2\text{-}HO\text{-}5\text{-}(CH_3)_3CC_6H_3)_2S$	$C_6H_5(CH_3)_2COOH$	353–383	7.85 – 50.7/θ	16
346	Di-(2-hydroxy-5-methyl)phenyl sulfide; $(2\text{-}HO\text{-}5\text{-}CH_3C_6H_3)_2S$	$C_6H_5(CH_3)_2COOH$	353–383	11.11 – 79.0/θ	16
347	Di-(2-hydroxy)phenyl disulfide; $(2\text{-}HOC_6H_4)_2S_2$	$C_6H_5(CH_3)_2COOH$	353–383	8.50 – 52.8/θ	16
348	2-Hydroxybenzyl-2-hydroxy-5-*tert*-butylphenylsulfide; $2\text{-}HO\text{-}5\text{-}(CH_3)_3CC_6H_3SCH_2(2\text{-}HOC_6H_4)$	$C_6H_5(CH_3)_2COOH$	343–383	5.20 – 34.3/θ	13
349	2-Hydroxy-5-methylbenzyldiethyl thiuram; $2\text{-}HO\text{-}5\text{-}CH_3\text{-}C_6H_3CH_2C(S)N(C_2H_5)_2$	$C_6H_5(CH_3)_2COOH$	343–383	14.60 – 98.7/θ	13
350	Phenylthiuram-N-benzyl; $C_6H_5CSNHCH_2C_6H_5$	$C_6H_5(CH_3)_2COOH$	343–383	15.40 – 97.6/θ	13
351	Thiobenzoyl-N-morpholine; *cyclo*-$[(CH_2)_2O(CH_2)_2N]C(S)C_6H_5$	$C_6H_5(CH_3)_2COOH$	343–383	7.70 – 46.6/θ	13

Table 7.6
Rate constants of trimolecular catalytic decomposition of hydroperoxides under action of products of oxidation of S-containing compounds in chlorobenzene

No.	S-Compound	Hydroperoxide	T/ K	k/ $l^2 mol^{-2} s^{-1}$ or $\log k = A - E/\theta$	Ref.
	$v = k\,[RSR][ROOH]^2$				
352	2-Hydroxybenzyl-*n*-pentyl sulfide; $2\text{-}HOC_6H_4CH_2SC_5H_{11}$	$C_6H_5(CH_3)_2COOH$	—	11.00 – 71.0/θ	17
353	2-Hydroxy-5-*tert*-butylphenylbenzyl sulfide; $2\text{-}HO\text{-}5\text{-}(CH_3)_3CC_6H_3SC_6H_5$	$C_6H_5(CH_3)_2COOH$	343–383	7.40 – 47.4/θ	13
354	2-Hydroxy-5-*tert*-butylphenylmethyl sulfide; $2\text{-}HO\text{-}5\text{-}(CH_3)_3CC_6H_3SCH_3$	$C_6H_5(CH_3)_2COOH$	383	0.59	17
355	2-Hydroxy-5-*tert*-butylthiophenol; $2\text{-}HO\text{-}5\text{-}(CH_3)_3CC_6H_3SH$	$C_6H_5(CH_3)_2COOH$	—	3.98 – 19.6/θ	17
356	2-Hydroxy-1,3-dithiophenol; $2,6\text{-}(HS)_2C_6H_3OH$	$C_6H_5(CH_3)_2COOH$	343–383	11.80 – 66.5/θ	13
357	2-Hydroxy-5-methylbenzyl-*n*-heptylsulfide; $2\text{-}HO\text{-}5\text{-}CH_3C_6H_3CH_2SC_7H_{15}$	$C_6H_5(CH_3)_2COOH$	—	11.60 – 78.6/θ	17
358	2-Hydroxy-5-methylbenzylphenyl sulfide; $2\text{-}HO\text{-}5\text{-}CH_3C_6H_3CH_2SC_6H_5$	$C_6H_5(CH_3)_2COOH$	—	9.58 – 72.7/θ	17

Table 7.7
Rate constants of reaction of peroxyl radicals with metal carbamates $[R_2NS_2]_2Me$ in hydrocarbon solutions

Metal	R	Peroxyl radical	T/ K	k/ $l\ mol^{-1}\ s^{-1}$	Ref.
Zn	C_2H_5	$C_6H_5(CH_3)CHO_2^•$	348	2.1×10^3	18
Pb	C_2H_5	$C_6H_5(CH_3)CHO_2^•$	348	9.0×10^3	18
Ni	C_2H_5	$C_6H_5(CH_3)CHO_2^•$	348	2.8×10^4	18
Cu	C_2H_5	$C_6H_5(CH_3)CHO_2^•$	348	1.0×10^5	18
Bi	C_2H_5	$C_6H_5(CH_3)CHO_2^•$	348	1.4×10^4	18
Ni	C_2H_5	$C_6H_5(CH_3)_2CO_2^•$	333	2.3×10^3	19
Zn	C_2H_5	$1\text{-}C_{10}H_{11}O_2^•$	303	1.3×10^2	20
Ni	$(CH_3)_2CH$	$1\text{-}C_{10}H_{11}O_2^•$	323	6.0×10^3	20
Ni	$(CH_3)_2CH$	$C_6H_5CH(O_2^•)CH_2\sim$	323	9.0×10^2	20
Ni	$(CH_3)_2CH$	$HO_2^•$	303	7.0×10^4	20
Ni	$CH_3(CH_2)_3$	$1\text{-}C_{10}H_{11}O_2^•$	323	1.4×10^4	20
Ni	$CH_3(CH_2)_3$	$C_6H_5CH(O_2^•)CH_2\sim$	323	5.0×10^3	20
Zn	$CH_3(CH_2)_3$	$C_6H_5(CH_3)_2CO_2^•$	333	2.4×10^3	21
Ni	$CH_3(CH_2)_3$	$C_6H_5(CH_3)_2CO_2^•$	333	1.9×10^3	19
Zn	$CH_3(CH_2)_3$	$C_6H_5(CH_3)_2CO_2^•$	333	4.7×10^3	22
Sb	$CH_3(CH_2)_4$	$C_6H_5(CH_3)_2CO_2^•$	333	3.8×10^3	22
Zn	$(CH_3)_3CCH_2CH(CH_3)CH_2$	$C_6H_5(CH_3)_2CO_2^•$	333	6.1×10^3	22
Sb	$(CH_3)_3CCH_2CH(CH_3)CH_2$	$C_6H_5(CH_3)_2CO_2^•$	333	3.5×10^3	22
Zn	*cyclo*-C_4H_7	$C_6H_5(CH_3)_2CO_2^•$	333	4.5×10^3	22
Zn	*cyclo*-C_6H_{11}	$C_6H_5(CH_3)_2CO_2^•$	333	3.5×10^3	22
Zn	*cyclo*-$[CH_2CH(-C_8H_{17})CH_2CH]$	$C_6H_5(CH_3)_2CO_2^•$	333	4.6×10^3	22
Zn	*cyclo*-$C_{12}H_{23}$	$C_6H_5(CH_3)_2CO_2^•$	333	8.0×10^2	22
Zn	*cyclo*-$[CH_2CH(-C_{18}H_{37})CH_2CH]$	$C_6H_5(CH_3)_2CO_2^•$	333	4.0×10^2	22

Table 7.8
Rate constants of reaction of metalthiophosphates and metalthiocarbamates with cumylhydroperoxide

Metal complex	T/ K	k/ l mol^{-1} s^{-1} or log k = $A - E/\theta$	Ref.
$[((CH_3)_2CHO]_2PS_2)_2Ni$	298–353	$7.55 - 55.5/\theta$	23
$[((CH_3)_2CHO]_2PS_2)_2Ni$	303	4.0×10^{-3}	24
$[((CH_3)_2CHO]_2PS_2)_2Zn$	328	3.8×10^{-3}	25
$(4\text{-}(CH_3)_3CC_6H_4O)_2PS_2)_2Ni$	298–353	$4.00 - 39.0/\theta$	23
$[(C_2H_5)_2NCS_2]_2Ni$	293–358	$2.70 - 29.9/\theta$	19
$[((CH_3)_2CHO]_2NCS_2]_2Ni$	303	0.16	24
$[(C_4H_9)_2NCS_2]_2Zn$	318–358	$6.06 - 54.0/\theta$	21
$[(C_4H_9)_2NCS_2]_2Ni$	293–358	$2.72 - 20.9/\theta$	19
$[(CH_3)_3CCH_2(CH_3)CHNCS_2]_2Zn$	293–358	$4.92 - 45.7/\theta$	21

REFERENCES

1 **Schwetlick, K., Konig, T., Ruger, C., Pionteck, J., Habicher, W. D.**, Chain-Breaking antioxidant activity of phosphite esters, *Polym. Degrad. Stab.*, 15, 97, 1986.

2 **Furimsky, E., Howard, J.A.**, Absolute rate constants for the reaction of *tert*-butylperoxy radicals with trivalent phosphorus compounds, *J. Am. Chem. Soc.*, 95, 369, 1973.

3 **Ruger, C., Konig, T., Schwetlick, K.**, Kinetik und Mechanismus der Cumylhydroperoxid zersetzung durch cyclische phosphite, *J. Prakt. Chem.*, 326, 622, 1984.

4 **Chebotareva, E. G., Pobedimskii, D. G., Kolubakina, N. S., Mukmeneva, N. A., Kirpichnikov P. A., Akhmadullina, A. G.**, Kinetics of reaction of cumylhydroperoxide with phosphites, *Kinet. Katal.*, 14, 891, 1973.

5 **Schwetlick, K., Ruger, C., Noack, R.**, Organophosphorus antioxidants. 1. Kinetics and mechanism of the decomposition of alkylhydroperoxides by *o*-phenylene phosphites and phosphates, *J. Prakt. Chem.*, 324, 697, 1982.

6 **Hiatt, R., Smithe, R. J., McColeman, C.**, The reaction of hydroperoxides with triphenylphosphine, *Can. J. Chem.*, 49, 1707, 1971.

7 **Aslanov, A. D.**, *Organic disulfides as inhibitors of oxidative processes*, Cand. Sci. (Chem.) Thesis Dissertation, Inst. Chem. Phys., Chernogolovka, 1985 (in Russian).

8 **Howard, J. A., Korcek, S.**, Absolute rate constants for hydrocarbon antioxidation. Part 20. Oxidation of some organic sulfides, *Can. J. Chem.*, 49, 2178, 1971.

9 **Kennerly, G. W., Patterson, W. L.**, Decomposition of cumene hydroperoxide in white mineral oil at 150 C by sulfur compounds, *Ind. Eng. Chem.*, 48, 1917, 1956.

10 **Koelewijn, P., Berger, H.**, Mechanism of the antioxidant action of dialkyl sulfoxides, *Rec. Trav. Chim.*, 91, 1275, 1972.

11 **Bridgewater, A. J., Sexton, M. D.**, Mechanism of antioxidant action: reactions of alkyl and aryl sulfides with hydroperoxides, *J. Chem. Soc. Perkin Trans. II*, 1978, 530.

12 **Zolotova, N. V., Gervitz, L. L., Denisov, E. T.,** Reactions of cumylhydroperoxide and cumylperoxyl radical with sulfides, *Neftekhimiya,* 15, 146, 1975.

13 **Kuliev, F. A.,** *Organic sulfides, disulfides and thioamides as inhibitors of oxidation,* Doct. Sci. (Chem.) Thesis Dissertation, Petrochem. Inst., Ufa, 1986 (in Russian).

14 **Aslanov, A. D., Zolotova, N. V., Denisov, E. T., Kuliev, F. A.,** Kinetics of reaction of hydroperoxides with some disulfides, *Neftekhimiya,* 22, 504, 1982.

15 **Akhundova, M. M., Farzaliev, V. M., Solyanikov, V. M., Denisov, E. T.,** Catalytic decomposition of cumylhydroperoxide by aminesulfides and inhibiting activity of products of oxidation of aminesulfides, *Izv. Akad Nauk SSSR, Ser. Khim.,* 1981, 741.

16 **Farzaliev, V. M., Kuliev, F. A., Akhundova, M. M., Denisov, E. T.,** The study of kinetics and mechanism of inhibiting action of sulfur-containing alkylated phenols, *Azerb. Khim. Zh.,* 1981, 117.

17 **Aliev, A. S., Farzaliev, V. M., Abdullaeva, F. A., Denisov, E. T.,** Mechanism of inhibiting action of oxyphenylsulfides on cumene oxidation, *Neftekhimiya,* 15, 890, 1975.

18 **Mazaletskii, A. B.,** *The antioxidant activity of S-containing chelates of heavy metals in oxidizing hydrocarbons,* Cand. Sci. (Chem.) Thesis Dissertation, Inst. Chem. Phys., Chernogolovka, 1979, Chap. 4 (in Russian).

19 **Gervits, L. L., Zolotova, N. V., Denisov, E. T.,** Mechanism of inhibiting action of dialkyldithiocarbamats of nickel in oxidizing cumene, *Neftekhimiya,* 15, 135, 1975.

20 **Howard, J. A., Chenier, J. H. B.,** Metal complexes as antioxidants. 2. Reaction of nickel dialkyldithiophoshates with alkylperoxy radicals, *Can. J. Chem.,* 54, 382, 1976.

21 **Korenevskaya, R. G., Kuzmina, G. N., Markova, E. I., Sanin, P. I.,** Kinetics of reaction of zinkdialkyldithiocarbamates with peroxyl radicals and cumylhydroperoxide, *Neftekhimiya,* 22, 477, 1982.

22 **Shelkova, R. G.,** *Mechanism action of dialkyldithiocarbamates of metals as effective inhibitors of hydrocarbon oxidation,* Cand. Sci. (Chem.) Thesis Dissertation, Petrochem. Inst., Moscow, 1990 (in Russian).

23 **Edilashvili, I. L., Ioseliani, K. B., Zolotova, N. V., Bakhturidze, G. Sh., Dzhanibekov, N. F.,** The study of mechanism of inhibiting action of diisopropyldithiophosphate and *o,o*-di-*para*-*tert*-butylphenyldithiophospate of nickel in oxidizing cumene, *Neftekhimiya,* 29, 892, 1979.

24 **Howard, J. A., Chenier, J. H. B.,** Metal complexes as antioxidants. 3. Reaction of nickel dialkyldithiocarbamates and nickel dialkyldithiophospates with alkyl hydroperoxides, *Can. J. Chem.,* 54, 390, 976.

25 **Shopov, D. M., Ivanov, S. K.,** *Reaction mechanism of inhibitors of peroxide decomposers.* Publishing house of the Bulgarian Academy of Sciences, Sofia, 1988, Chap. 2.

INDEX

V

Printed and bound by CPI Group (UK) Ltd, Croydon, CR0 4YY

22/10/2024

01777630-0020